Lecture Notes in Computer Science   1426
Edited by G. Goos, J. Hartmanis and J. van Leeuwen

Lecture Notes in Computer Science 1426

Edited by G. Goos, J. Hartmanis and J. van Leeuwen

Springer-Verlag Berlin Heidelberg Gmbh

Frédéric Geurts

Abstract Compositional Analysis of Iterated Relations

A Structural Approach
to Complex State Transition Systems

 Springer

Series Editors

Gerhard Goos, Karlsruhe University, Germany
Juris Hartmanis, Cornell University, NY, USA
Jan van Leeuwen, Utrecht University, The Netherlands

Author

Frédéric Geurts
Service de Mathématiques de la Gestion
Université Libre de Bruxelles
CP 210 01, Boulevard du Triomphe
B-1050 Bruxelles, Belgium
E-mail: fgeurts@smg.ulb.ac.be

Cataloging-in-Publication data applied for

Die Deutsche Bibliothek - CIP-Einheitsaufnahme

Geurts, Frédéric:
Abstract compositional analysis of iterated relations : a structural approach
to complex state transition systems / Frédéric Geurts. - Berlin ; Heidelberg
; New York ; Barcelona ; Hong Kong ; London ; Milan ; Paris ; Singapore ; Tokyo
: Springer, 1998
   (Lecture notes in computer science ; Vol. 1426)

CR Subject Classification (1998): F.1, F.3.1, C.3, D.2.4

ISSN 0302-9743
ISBN 978-3-540-65506-0   ISBN 978-3-540-49211-5 (eBook)
DOI 10.1007/978-3-540-49211-5

© Springer-Verlag Berlin Heidelberg 1998
originally published by Springer-Verlag Berlin Heidelberg New York in 1998.

Typesetting: Camera-ready by author
SPIN 10637524   06/3142 – 5 4 3 2 1 0   Printed on acid-free paper

Foreword by Michel Sintzoff

The present book apparently falls outside of the scope of the LNCS series: the theory of dynamical systems is mainly used for systems defined by, say, differential equations, and very little for programs. Yet, to consider programs as dynamical systems sheds light at least on the relationship between discrete-time systems and continuous-time ones; this is an important issue in the area of hybrid systems, where control engineers and software designers learned to work hand in hand.

As a matter of fact, program traces constitute time-to-state functions, and programs which define sets of traces characterize reactive systems as used in industry and services. Quite similarly, differential systems define sets of time-to-state functions, and they serve in many disciplines, e.g. physics, engineering, biology, and economics. Thus, we must relate programs as well as differential equations to dynamical systems.

The concepts of invariance and attraction are central to the understanding of dynamical systems. In the case of programs, we use the quite similar notions of invariance, viz. safety, and reachability, viz. termination or liveness; reachability amounts to finite-time attraction and weakest preconditions determine largest basins of reachability. Accordingly, the basic programming concepts of fairness, fault-tolerance and self-stabilization correspond, in the case of dynamical systems, to recurrence (repeated return to desired states), structural stability (return to desired dynamics after system perturbation), and absorption (return to a desired invariant after state perturbation).

Linear dynamical systems are usually analyzed in terms of analytical expressions which provide explicit solutions for simple differential or difference equations. In the case of nonlinear dynamical systems, exact solutions cannot be obtained in general, and the qualitative analysis is then carried out on the system specifications themselves, viz. on differential equations. For instance, attraction is proven using an energy-like function: the successive dynamical states are abstracted to decreasing non-negative reals. Also, the qualitative analysis of concrete dynamics can be reduced to that of symbolic ones, in which each state is a symbol abstracting a set of concrete states; this shows discrete dynamics can serve as qualitative abstractions of continuous ones.

Similarly to nonlinear systems, programs in general cannot be understood in terms of analytical solutions. Weakest preconditions often become too complex, and practical reasoning methods apply on the programs themselves. For example, invariance is checked by structural induction, and termination is verified using an energy-like function from the successive dynamical states to decreasing non-negative integers. Moreover, the verification of a concrete program, very much as in the case of a concrete nonlinear dynamical system, is better carried out in terms of an abstract, simpler one. This paradigm of abstraction underlies many useful techniques in mathematics as well as in computing; let us recall automata simulation, data representation, abstract interpretation, and time abstraction.

Interestingly enough, the mathematical theory of dynamical systems not only supports abstraction-based methods, e.g. symbolic dynamics, but also introduces basic compositional techniques such as sequential and iterative composition. What could then computing science contribute to that theory? The answer is clear: *scaling up*. Actually, the central results in the classical theory of dynamical systems concern single-level individual systems. For us, the main challenge is to design systems for many complementary goals and at various abstraction levels. To this end, we intensively use the principles of modular composition and stepwise refinement. The same approach could give rise to possible original contributions of computing science in the area of dynamical systems. Indeed, the present book shows how to construct complex dynamics by a systematic composition of simple ones, and thus provides a roadmap to compositional design techniques for scaled-up dynamical systems.

Programming theory has taken great advantage of logic and algebra. It should similarly benefit from the theory of dynamical systems; this synergy would entail a common scientific platform for system engineering at large, including software engineering. Examples of such cross-fertilization already exist. Discrete-event control systems and hybrid systems, combining continuous and discrete time, are specified, analyzed, and synthesized using finite-state automata. Synchronization of dynamics provides a means of secure communication. Emergent computations can be implemented by cellular neural networks. Distributed dynamics help to analyze agent-based systems.

The nice matching between dynamics and computational intuitions explains the success of automata-based requirements, dynamics-based architectures, state-based specifications, object-oriented systems, proof dynamics, and design-process models. At each abstraction level, dynamics can be specified at will using programs, automata, logic, algebra, or calculus. For many-sided and multi-level systems such as the web or a house, the crucial issues are the choice of the right level of dynamics, the interaction of internal dynamics with partially defined external ones, and the scaling-up of state-, control- and time-refinements.

The author must be thanked warmly for providing us with many stimulating ideas on these attractive themes.

Preface

State-transition systems model machines, programs, and specifications [20, 23, 284, 329], but also the growth and decline of ant populations, financial markets, diseases and crystals [22, 35, 178, 209, 279]. In the last decade, the growing use of digital controllers in various environments has entailed the convergence of control theory and real-time systems toward hybrid systems [16] by combining both discrete-event facets of reality with Nature's continuous-time aspects. The computing scientist and the mathematician have re-discovered each other. Indeed, in the late sixties, the programming language Simula, "father" of modern object-oriented languages, had already been specifically designed to model dynamical systems [76].

Today, the importance of computer-based systems in banks, telecommunication systems, TVs, planes and cars results in larger and increasingly complex models. Two techniques had to be developed and are now fruitfully used to keep analytic and synthetic processes feasible: composition and abstraction. A compositional approach builds systems by composing subsystems that are smaller and more easily understood or built. Abstraction simplifies unimportant matters and puts the emphasis on crucial parameters of systems.

In order to deal with the complexity of some state-transition systems and to better understand complex or chaotic phenomena emerging out of the behavior of some dynamical systems, the aim of this monograph is to present first steps toward the integrated study of composition and abstraction in dynamical systems defined by iterated relations.

The main insights and results of this work concern a structural form of complexity obtained by composition of simple interacting systems presenting opposed attracting behaviors. This complexity expresses itself in the evolution of composed systems, i.e., their dynamics, and in the relations between their initial and final states, i.e., the computations they realize. The theoretical results presented in the monograph are then validated by the analysis of dynamical and computational properties of low-dimensional prototypes of chaotic systems (e.g. Smale horseshoe map, Cantor relation, logistic map), high-dimensional spatiotemporally complex systems (e.g. cellular automata), and formal systems (e.g. paperfoldings, Turing machines).

Acknowledgements. This monograph is a revision of my PhD thesis which was completed at the Université catholique de Louvain (Belgium) in March 96.

The results presented here have been influenced by many people and I would like to take this opportunity to thank them all.

In particular, I express my deepest gratefulness to my advisor, Michel Sintzoff, with whom I had the rewarding privilege to collaborate. His generous support, his never-ending interest in my work, his incredibly long-term scientific perspective, and his matchless sense of humour incited me to develop and write things I would never have dreamt of. I owe Michel an abstract compositional virus that flies in the Garden of Structural Similarities.

My gratitude further goes to Yves Kamp, André Arnold, and Michel Verleysen, for their careful reading of draft versions of this text, and for their kind and constructive way to turn simple statements into convincing ones. I am also thankful to Nicola Santoro and Paola Flocchini for their constant belief in my research on cellular automata, and for their multiple invitations to Ottawa.

I wish to thank the staff of the Computer Science Department at UCL, and especially its chairman, Elie Milgrom, for providing the nice environment in which I could spend five exciting years.

I acknowledge the financial support I received from the *Fonds National de la Recherche Scientifique*, the *European Community*, the *Communauté Française de Belgique*, and the *Académie Royale de Belgique*.

My warmest thanks go to my friends Bruno Charlier, Luc Meurant, and Luc Onana Alima, for their irreplaceable presence, and to my parents and sister, Pol, Rose-Marie, and Muriel, for their eternal love, care, and attention.

At last but not least, words are not strong enough to tell my love to my wife, Cécile, and to our beautiful smiling daughter, Romane. Without their emotional support, all this would not have been possible.

> I had a dream.
> I was there, under the sun,
> Waiting for nothing, for happiness.
> Quelque chose attira mon attention.
> Etait-ce cet oiseau qui volait vers moi ?
> Il y avait tant de monde que j'avais peine à distinguer
> D'où venait cette douce magie qui m'enrobait.
> Puis des notes, une musique sublime, se dévoilèrent,
> Et tu apparus, Vénus, d'un océan de joie,
> Enivrant de ta douceur bleue le ciel et tous ses astres.

Frédéric Geurts
Louvain-la-Neuve, Belgium
January 1998

Table of Contents

Part II. Abstract Complexity: Abstraction, Invariance, Attraction

---

Part III. Abstract Compositional Analysis of Systems: Dynamics and Computations

1. Prologue: Aims, Themes, and Motivations

> L'ordre est le plaisir de la raison
> mais le désordre est le délice de l'imagination.
>
> Paul Claudel

> La vie est pour chaque homme surgi du chaos
> une goutte d'eau douce entre deux océans.
> Cette tendresse au coeur, nous roulons vers la mer
> et notre soif s'enflamme au sel d'éternité.
>
> Henri Coppieters de Gibson

The paradigms of abstraction and composition are central in many areas of computer science and mathematics. For instance, important results concern program refinement and construction, parallel and distributed systems, abstract data types, proof checking, VLSI circuits, and artificial neural networks.

We strongly believe that the same tools can be fruitfully extended to a more general setting including chaotic systems, mathematical models of state-based transition systems, parallel programs and cooperative agents, and thereby enable a better understanding of the wide variety of complex phenomena they exhibit.

The goal and main contribution of this monograph is to present first developments in the abstract compositional analysis of dynamical and computational properties of discrete-time relational dynamical systems. We study properties of composed systems by combining the individual analyses of their components, together with abstraction techniques. This allows us to propose a structural view of dynamical complexity, as well as a structural computational hierarchy of dynamical systems.

This introductory chapter is organized as follows: §1.1 illustrates the context of our study by means of examples; §1.2 presents and motivates the main tools used within this context; finally, §1.3 gives an overview of the monograph.

F. Geurts: Abstract Compositional Analysis of Iterated Relations, LNCS 1426, pp. 1-18, 1998.

1.1 Complex Relational Dynamical Systems

This section presents dynamical systems based on iterated relations, first in a semi-theoretic way, then by means of paradigmatic examples of computing science and mathematitcs which illustrate the main properties and concepts introduced and developed in the next chapters: first, a classical mutual exclusion algorithm; then, a very simple model of social pressure; finally, the well-known chaotic evolution of rabbit populations. These examples also show the wide range of applications in which dynamical systems can be used as models and explanatory tools of natural and artificial phenomena.

1.1.1 The Context: A First Contact with Dynamical Systems

Iterative Dynamical Systems. Dynamical systems are abstract mathematical objects [76] describing the time evolution of concrete objects like the atmospheric pressure [207], the biodiversity of an ecosystem [31], the temperature of a nuclear reactor [16], the load level of a computer network, exchange rates between US dollars and Belgian Francs or other standard indicators of financial markets [22, 82], etc. These concrete objects can be in several states, generally obtained by means of physical measures. Any discrete-time evolution is a transition between states. Thus, if a concrete object is in a given state, the next state following a transition will characterize the object after a certain amount of time. When time is discrete, transitions are obtained by application of a function defined on the state space of the object; they can be expressed by difference equations or by programs. When continuous-time evolutions are considered instead, transitions become infinitesimal in time and are expressed by differential equations. In this monograph, we consider the discrete-time level only.

Let us thus introduce some "soft" mathematics. Let C be the concrete object, m be an "instrument" measuring the state of C at a given time: m is a function from time to the space X that contains all possible states of C. The dynamical system providing the discrete-time evolution of C is a pair (X, f) where f is a function from and to X. Thus, if $m(t) = x$ then $m(t + 1) = f(x)$, where time is normalized in such a way the increment between two observations is always 1. A longer evolution can be obtained as follows: let C be in state x_0 at time 0, viz. $m(0) = x_0$. The evolution from x_0 results from successive iterations as follows:

$$\underbrace{x_0}_{m(0)} \xrightarrow{f} \underbrace{f(x_0)}_{m(1)} \xrightarrow{f} \underbrace{f(f(x_0))}_{m(2)} \xrightarrow{f} \underbrace{f(f(f(x_0)))}_{m(3)} \xrightarrow{f} \cdots$$

The evolution from a set of states A_0 is obtained by successive iterations

$$A_0 \xrightarrow{f} f(A_0) \xrightarrow{f} f(f(A_0)) \xrightarrow{f} f(f(f(A_0))) \xrightarrow{f} \cdots$$

of the set-transformer, or point-to-set lifting, of $f \colon \forall A \subseteq X$,

$$f(A) = \cup_{a \in A} f(a).$$

The full dynamics of the system is a set that contains all possible evolutions, from all possible starting points: one evolution per initial state.

Relations as Nondeterministic Dynamical Systems. In order to model concrete problems, for example in computing science, nondeterminism has to be added to the notion of dynamical systems, and functions must be extended to relations. Indeed, nondeterminism plays a crucial role in models of asynchronous parallelism, and in relational aspects of logic programming or databases. There is also a theoretical reason for preferring relations to functions, namely to provide a homogeneous mathematical framework for the study of dynamical systems.

When a relation is considered to describe transitions between states of a specific system, several transitions can occur from a given state and lead to new states which, in turn, can lead to several next states, etc. For instance, f being a relation defined on X, for a given state x, there can be several images y such that $(x, y) \in f$. Each nondeterministic transition can be denoted by

$$x \xrightarrow{f} y$$

but we can also regard the relation as a multi-valued function, and take all possible images as next "state"

$$x \xrightarrow{f} \{y \mid (x, y) \in f\}.$$

This immediately requires to introduce set-transformers on top of multi-valued functions because we want to compute a complete second iteration from x, that is, the image of a set.

Here, fixing one initial state x_0 is no more equivalent to giving one possible evolution from x_0, because of nondeterminism: different evolutions can correspond to a single initial point. Still, however, the full dynamics can be defined as the set of all possible evolutions from all possible starting states.

Complexity in Dynamical Systems. There are two ways to look at systems and their evolution. One can look at the evolution itself, i.e. the dynamics of the system, or one can concentrate on the relations between initial and final or asymptotic states, i.e. the computations of the system.

The first aspect is classical in mathematics, physics, and program theory: chaos, ergodicity, fractals, but also invariance, termination, fairness, and self-stabilization, all these notions describe *how systems visit their underlying spaces as they evolve.* The second aspect comes from computability theory, where one looks at *input-output relations* of automata, machines, programs, to define their *computational power.*

The remaining subsections illustrate these notions of evolution and complexity by means of classical examples of computer science and mathematics.

1.1.2 Mutual Exclusion

When two serious users of a computer network want to print the results of looong hours of labour, they probably prefer them to be separated and not mixed! To make sure that their wish becomes reality, the two underlying user processes have to talk to the printer-management process sequentially, and not together; the user processes must have a mutually exclusive access to the printer resource. In [262], Peterson proposed a very simple algorithm solving the mutual exclusion problem. Our aim here is not to prove that the algorithm is correct, but rather to show an abstract way of analyzing its behavior, and showing that it fulfills its requirements.

Peterson's Algorithm. Let us first present a guarded-command-like [91] version of Peterson's algorithm for two processes whose purpose is to guarantee a sequentialized execution of a "critical section" from which a given resource must be accessed in a mutually exclusive way. The three components (initialization part plus two core processes) and their parallel composition are defined in Fig. 1.1.

$$
\begin{array}{rcl}
I & = & d_0 := \mathbf{f}; d_1 := \mathbf{f}; t := 0 \\[2mm]
P_{0/1} & = & \text{do } \mathbf{t} \rightarrow \quad \begin{array}{ll}
(1) & \text{non critical section;} \\
(2) & d_{0/1} := \mathbf{t}; \\
(3) & t := 0/1; \\
(4) & \text{do } (d_{1/0} = \mathbf{t} \wedge t = 0/1) \rightarrow \text{ skip od;} \\
(5) & \text{critical section;} \\
(6) & d_{0/1} := \mathbf{f}
\end{array} \\
& & \quad \text{od} \\[2mm]
P & = & I; (P_0 \parallel P_1)
\end{array}
$$

Fig. 1.1. Peterson's mutual exclusion algorithm

The parallel composition $P_0 \parallel P_1$ can be executed synchronously or not. In the first case, the instructions or transitions of both processes are executed at the same time. In the second case, a nondeterministic choice determines which process executes its transition, and a fair scheduling [110] prevents any one of them from being discarded forever.

Before analyzing the behavior of P, let us rewrite the main components P_0 and P_1 as state-transition systems (see Fig. 1.2).

Asynchronous Execution. Let us start with the asynchronous version, which seems more realistic as a global clock might not always be available to permit a perfect synchronization. Table 1.1 shows all possible transitions. Each cell of this table contains two lines, respectively denoting a transition of S_1 (top) or a transition of S_0 (bottom). Some transitions are disabled, in which case the corresponding half cells are left empty.

$$S_{0/1} = \{ \quad \begin{aligned} &1 \vdash \text{non critical section} \mapsto 2, \\ &2 \vdash d_{0/1} := \mathbf{t} \mapsto 3, \\ &3 \vdash t := 0/1 \mapsto 4, \\ &4 \vdash d_{1/0} = \mathbf{f}? \mapsto 5, \\ &4 \vdash t = 1/0? \mapsto 5, \\ &5 \vdash \text{critical section} \mapsto 6, \\ &6 \vdash d_{0/1} := \mathbf{f} \mapsto 1 \end{aligned} \\ \}$$

Fig. 1.2. Peterson's mutual exclusion transition system

The crucial point is $(3,3)$, from which the choices $(3,4)$ and $(4,3)$ entail two different chains of transitions: either

$$(3,3) \to (3,4) \vdash t := 0 \mapsto (4,4) \to (4,5) \to (4,6) \to (4,1) \to (5,2)$$

or

$$(3,3) \to (4,3) \vdash t := 1 \mapsto (4,4) \to (5,4) \to (6,4) \to (1,4) \to (2,5).$$

By looking at the table, we notice that no cell contains the pair $(5,5)$. Good! It means that the mutual exclusion of the critical section is verified.

Table 1.1. Asynchronous transitions of Peterson's mutual exclusion algorithm

S_1

1	2	1	3	1	4	1	5	1	6	1	1
2	1	2	2	2	3	2	4	2	5	2	6
2	2	2	3	2	4	2	5	2	6	2	1
3	1	3	2	3	3	3	4	3	5	3	6
3	2	3	3	3	4			3	6	3	1
4	1	4	2	4	3	4	4	4	5	4	6
4	2	4	3	4	4	4	5	4	6	4	1
5	1	5	2			5	4				
5	2	5	3	5	4			5	6	5	1
6	1	6	2	6	3	6	4	6	5	6	6
6	2	6	3	6	4			6	6	6	1
1	1	1	2	1	3	1	4	1	5	1	6

S_0 (row label, at left of table)

Some other states are also unreachable as the only path toward them has to start from or contain $(5,5)$:

$$\{(5,5), (5,6), (6,5), (6,6)\}.$$

This set of configurations is called poetically *Garden of Eden*: transition paths from the Garden are one way only, no path ever reaches or returns to the Garden.

Synchronous Execution. Let us quickly review the synchronous execution of the algorithm. The new transition table is shown in Table 1.2.

Table 1.2. Synchronous transitions of Peterson's mutual exclusion algorithm

S_1

2	2	2	3	2	4	2	5	2	6	2	1
3	2	3	3	3	4	3	5	3	6	3	1
4	2	4	3	3	4	4	4	4	6	4	1
				4	3						
5	2	5	3	4	4	4	5	4	6	4	1
						5	4				
6	2	6	3	6	4	6	4	6	6	6	1
1	2	1	3	1	4	1	4	1	6	1	1

S_0 (row label on the left)

In this model, the transition from $(3,3)$ has to be split into two transitions if write-atomicity on variable t holds. Actually, one can rewrite the transition as a nondeterministic coin tossing

$$(t := 0, t := 1)$$
$$\equiv \quad t := 0; t := 1 \,\square\, t := 1; t := 0$$
$$\equiv \quad t := \text{coin tossing}.$$

The execution path from $(1,1)$ is obviously much simpler than in the asynchronous case (see Fig. 1.3).

The first part of this evolution, above the line, is called *transient*; because of the nondeterministic choice from $(3,3)$, the transient has two possible outcomes and is thus nondeterminstic. The second part, below the line, is a *loop* or *cycle* of length six; there, every transition is deterministic: once in the cycle, the evolution is determined forever; the present fully determines the future. By definition, the cycle is an *invariant* set of states: once in this set, all subsequent evolutions remain in the cycle. Moreover, after a finite amount of transitions in the transient, the system enters and remains in the thereby *attracting* cycle. All other states eventually lead to this evolution and constitute the Garden of Eden.

Comments. The algorithm P whose abstract state-transition version has been studied, is nothing but a dynamical system obtained by successive iterations of an underlying relation between global states composed of substates of each of the individual processes. Several important notions have already been illustrated in this first example. Two types of composition have been used to assemble the building blocks of P (sequential composition and parallel composition). We have used nondeterministic transitions in order to model some concrete aspects of the system. Two dynamical properties have been introduced: invariance and attraction.

We do not pretend that dynamical systems are the key to mutual exclusion algorithms. However, their use might be appropriate to analyze and understand the behavior of specific algorithms or to conceive ones verifying given properties.

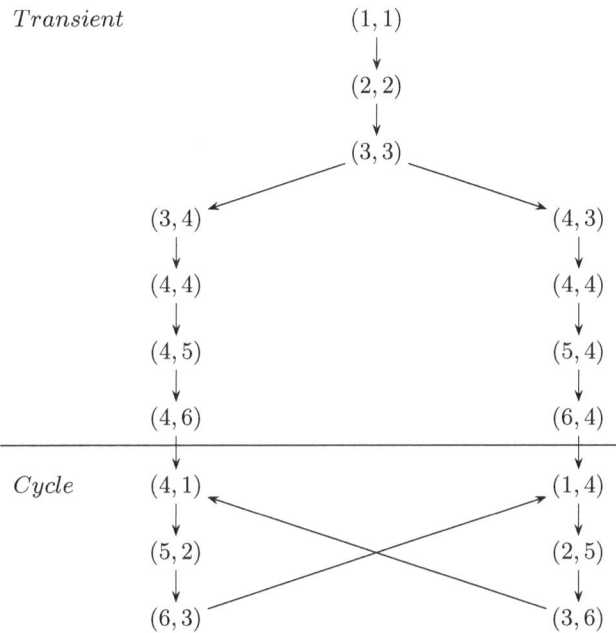

Fig. 1.3. Dynamics of the synchronous version of Peterson's algorithm

1.1.3 Social Pressure

Before election days and referenda, newspaper headlines often read "Polling results – 60% of the people think yellow (let us not make it controversial)". What are the other 40% of this population going to think after this publication? When you know that almost everybody around you thinks yellow, are you not tempted to go yellow? Your first answer is probably "No!". Anyway, this kind of social pressure is a common tool used by media, if any, all around the world.

Although not directly related to media, crucial parts of control systems of nuclear plants or of planes, space shuttles and other flying objects follow the same rules. They are very often duplicated, triplicated or even quadruplicated, run in parallel, and a supposedly reliable Comparator compares the results and takes either the most frequent one or their average, depending on the application. This example is a strong form of fault-tolerance: up to half of the components can fail and still the global system can reach the right target.

Majority Vote System. To illustrate this social pressure, let us consider a loop of four very simple automata (see Fig. 1.4), running synchronously and trying

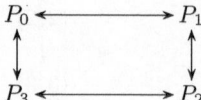

Fig. 1.4. Majority vote system: loop of four automata

to reach an agreement without any global comparison by taking a majority function of the values they observe in their neighborhood. The processes and their synchronous parallel composition are defined in Fig. 1.5.

$$
\begin{aligned}
P_i \;\; &= \;\; \textbf{do } X_{i\ominus 1} = X_{i\oplus 1} = 0 \rightarrow X_i := 0 \\
&\quad\;\; [\!] \; X_{i\ominus 1} \neq 0 \vee X_{i\oplus 1} \neq 0 \rightarrow X_i := 1 \\
&\quad\;\; \textbf{od}
\end{aligned}
$$

$$
P \;\; = \;\; P_0 \parallel P_1 \parallel P_2 \parallel P_3.
$$

Fig. 1.5. Majority vote system: algorithm

Each local algorithm P_i can also be described as in Table 1.3: the process compares the values of its neighborhood with all entries of the table, and it takes the corresponding value.

This last way of describing the behavior of automata is typical in the field of cellular automata, systems composed of a number of homogeneous automata computing their own values as simple functions of the values of their close neighbors [330].

Dynamics. What does the evolution of this group of automata look like? Since there are only sixteen different configurations, an exhaustive exploration is possible. If we denote a global configuration of the automata as a string of four binary digits, the evolution is represented in Fig. 1.6.

The dynamics is very simple. After one global step or iteration, the configuration of the system belongs to $\{0000, 1111, 1010, 0101\}$; the system is

Table 1.3. Majority vote system: tabular description of components

Old values			New value
$P_{i\ominus 1}$	P_i	$P_{i\oplus 1}$	P_i
0	0	0	0
0	0	1	1
0	1	0	0
0	1	1	1
1	0	0	1
1	0	1	1
1	1	0	1
1	1	1	1

thus attracted to this set of configurations. The *transient* is very short, as only one iteration is necessary to reach the *attractor*. Moreover, this set is strongly *invariant*: the system never leaves it, it either remains on 0000 or 1111 indefinitely, or it keeps oscillating between 1010 and 0101.

Comments and Bibliographic Notes. Again, this obvious example illustrates several important notions we will develop later on: a composition mechanism has been used to build a global automaton from four components and can be generalized to any finite or infinite composition; attraction and invariance have been used to describe the potential evolutions of the system.

Various versions of the social pressure dynamics have been studied in different contexts, from pure mathematics to immunology to distributed computing: [122, 267, 266] respectively studied the period-two-property of symmetric weighted majorities on finite $\{0, 1\}$-colored graphs, finite $\{0, \cdots, p\}$-colored graphs, and symmetric weighted convex functions on finite graphs; [7, 8, 124] examined the number of fixpoints of finite $\{0, 1\}$-colored rings; in [232, 231], the author analyzed strong majorities on finite and infinite $\{0, 1\}$-colored lines, and in [233], he further studied infinite connected $\{0, 1\}$-colored graphs. Dynamic majorities were applied to the immune system [7], in image processing [7, 121], and to fault-tolerant systolic arrays and VLSI circuits [240, 241, 242]. For more details, see also the recent survey [261], and the book [121].

1.1.4 On the Chaotic Demography of Rabbits

Imagine we want to describe the evolution of a population of rabbits, or the dynamics of another growth process such as an ecosystem or an economic market [22, 35, 175, 177, 209].

Linear Dynamics. If we model the population density at time t as a continuous value x_t of the interval $[0, 1]$, and the discrete evolution steps as a linear function with growth factor λ:

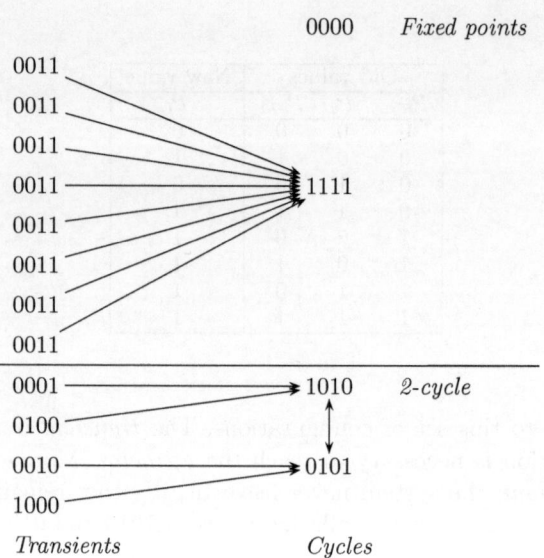

Fig. 1.6. Majority vote system: dynamics

$$x_{t+1} = \lambda x_t$$

the overall evolution is simply

$$x_t = \lambda^t x_0$$

starting from a level x_0 at time $t = 0$. Thus, the level asymptotically decreases to 0 if $0 < \lambda < 1$, it remains constant if $\lambda = 1$, and it explodes exponentially if $\lambda > 1$. The only fixed level is 0.

This first model is of course too trivial as it assumes the existence of a Paradise on Earth: no disease, enough food for every rabbit, no thermonuclear war between rabbit races, etc.

Nonlinear Dynamics. To make the model more plausible, let us thus replace the constant growth factor λ by a variable one, $\lambda(1 - x_t)$: the more rabbits, the more problems, the less rabbits at the next generation. This saturation factor taken into account, we get the so-called *logistic map*

$$x_{t+1} = f_\lambda(x_t) = \lambda(1 - x_t)x_t.$$

Now, depending on the value of λ, lots of different behaviors can emerge. Are there, for example, fixed populations? Yes, there are two possibly fixed levels:

$$x = f_\lambda(x) \Leftrightarrow x = 0 \text{ or } x = p_\lambda = 1 - \frac{1}{\lambda}.$$

Are there cycles? Yes, up to infinitely many cycles can coexist in the interval $[0, 1]$. Are these fixpoints and cycles attracting? Not always, it varies along the λ axis, as shown by the *bifurcation* diagrams depicted in Fig. 1.7.

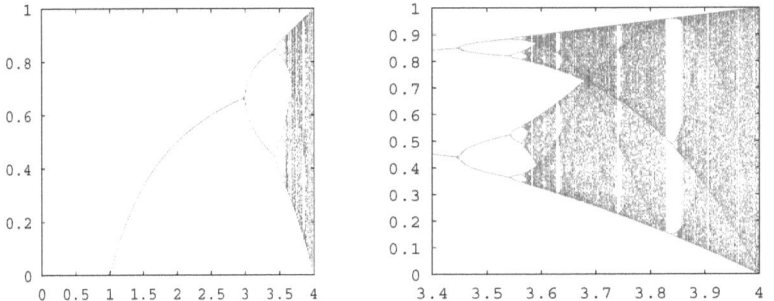

Fig. 1.7. Bifurcation diagrams of the logistic map: a random point is chosen and the system is iterated until attracted to a cycle, for different values of the parameter λ; the right-hand graph details a part of the left-hand one

Table 1.4 summarizes the effects of fixpoints, and Fig(s). 1.8-1.11 illustrate different behaviors corresponding to different values of the reproduction parameter λ. For each figure, the leftmost graph represents the function and successive iterations starting from a random point, and the rightmost graph represents the same evolution as a function of time.

Table 1.4. Local stability of the logistic map fixpoints 0 and p_λ

λ	$f'_\lambda(0)$	$|f'_\lambda(0)|$	Evolution
$(0, 1)$	> 0	< 1	Monotonic attraction
$(1, \infty)$	> 0	> 1	Monotonic repulsion

λ	$f'_\lambda(p_\lambda)$	$|f'_\lambda(p_\lambda)|$	Evolution
$(0, 1)$	> 0	> 1	Monotonic repulsion
$(1, 2)$	> 0	< 1	Monotonic attraction
$(2, 3)$	< 0	< 1	Oscillating attraction
$(3, \infty)$	< 0	> 1	Oscillating repulsion

All figures show that $[0, 1]$ is a global invariant. Figures 1.8-1.10 show that smaller fixpoint invariants can attract the global invariant asymptotically, whereas Figure 1.11 exhibits two important properties of complex dynamical systems. The leftmost graph shows on the function itself the evolution from a random point which does not seem to stabilize into a subinterval of $[0, 1]$ and keeps on visiting all of $[0, 1]$ in a scattered way. This property of

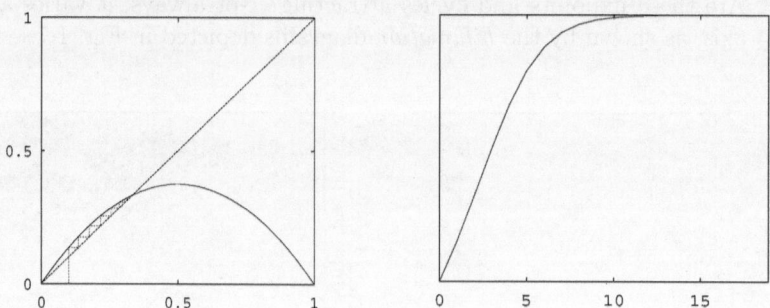

Fig. 1.8. Dynamics of $f_{1.5}$: one repelling fixpoint (0), and one attracting fixpoint ($p_{1.5}$)

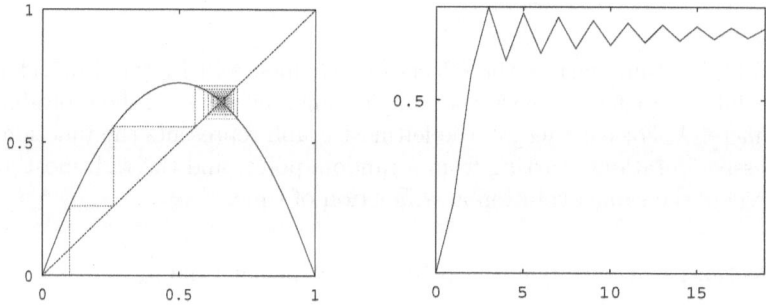

Fig. 1.9. Dynamics of $f_{2.9}$: one repelling fixpoint (0), and one attracting fixpoint ($p_{2.9}$)

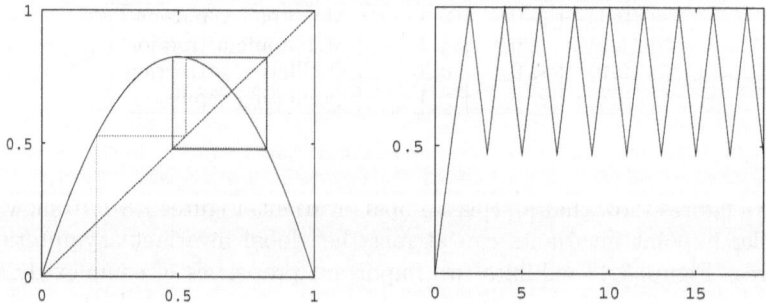

Fig. 1.10. Dynamics of $f_{3.3}$: two repelling fixpoints and one attracting 2-cycle

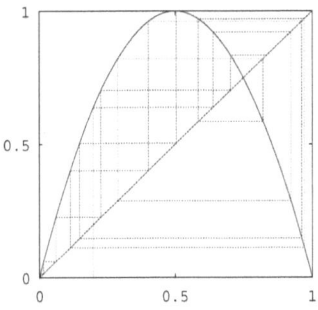 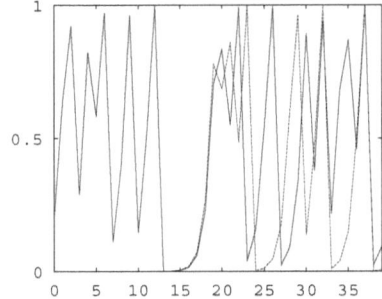

Fig. 1.11. Dynamics of f_4: chaos, infinitely many cycles, topological transitivity, sensitivity to initial conditions

irreducibility of the invariant is called *topological transitivity*. Its main symptom is the existence of points whose orbits never stabilize to subparts of the global invariant. The rightmost graph shows the same evolution plus another one starting from a very close initial point: after a couple of iterations, the two evolutions diverge. This illustrates a property called *sensitivity to initial conditions*: small initial perturbations can entail dramatically different evolutions. The conjunction of these two important properties constitute the essence of chaos.

Abstraction: Symbolic Dynamics. The complex behavior of f_4 can be analyzed in a very simple way, thanks to a well-known abstraction technique: symbolic dynamics. Roughly speaking, the abstraction consists in:

- dividing $[0, 1]$ into two subintervals $[0, \frac{1}{2}]$ and $(\frac{1}{2}, 1]$, respectively denoted by I_0 and I_1;
- mapping evolutions of points to symbolic sequences:

$$A \;:\; [0, 1] \mapsto \{0, 1\}^{\mathbb{N}}$$
$$\text{s.t.} \quad A(x) = s_0 s_1 s_2 \cdots \text{ where } \forall i, s_i = a \Leftrightarrow f^i(x) \in I_a;$$

- establishing a correspondence (i.e., a homomorphism) between the dynamics of f_4 and the shift dynamical system σ ($\sigma(s_0 s_1 s_2 \cdots) = s_1 s_2 \cdots$):

$$\forall x \in [0, 1], \sigma(A(x)) = A(f_4(x)).$$

The analysis of σ in its uncountably infinite space of sequences $\{0, 1\}^{\mathbb{N}}$ is straightforward; its main properties are:

- the existence of 2^n cycles of length n, for each n;
- the existence of a dense orbit, that is, an orbit which passes arbitrarily close to any state of $\{0, 1\}^{\mathbb{N}}$;
- the density of its periodic orbits in $\{0, 1\}^{\mathbb{N}}$;
- the sensitivity to initial conditions, on the space $\{0, 1\}^{\mathbb{N}}$ metrized by

$$d_a(x_0 x_1 x_2 \cdots, y_0 y_1 y_2 \cdots) = 2^{-\inf\{i | x_i \neq y_i\}}.$$

This abstraction based on the homomorphism A preserves topological properties, including the four ones listed above. Thus, we immediately get four results on the dynamics of f_4 without really analyzing f_4 itself, which shows the power of working out an appropriate abstraction between a given seemingly complex system and a simpler one.

Comments and Bibliographic Notes. The different values of the parameter λ give to f_λ a broad range of behaviors, from very simple (monotonic attraction to small invariants) to very complex (chaotic evolution in the global invariant).

The intent of this section was not to give a precise definition of chaos, which we will do in Chap. 5, but rather to emphasize the various levels of complexity that can be found in the structure of invariants and attractors.

The first form of logistic map was introduced in 1845 by the Belgian sociologist and mathematician P.-F. Verhulst to model the growth of populations limited by finite resources [311]. A very detailed study of the map can be found in [88]. The book [202] is entirely devoted to symbolic dynamics (see also Chap. 4).

1.2 Tools and Motivations

In the context of relational dynamical systems, we focus on the analysis of complex behaviors. Inspired by successful approaches in computing science and mathematics, we develop a *compositional analysis*, and we use it together with *abstraction* to study well-known families of systems.

Composition. Basically, the compositional analysis of a dynamical system consists in determining some global property concerning dynamical or computational aspects of the system by combination of individual properties of its components, which are expected to be simpler.

The following diagram illustrates the idea; S_i being the components of a composed system $S = \star_i S_i$, I denoting an individual property, and G a global property, we want to find a way to combine the individual properties, viz. $\diamond$, to characterize the global property:

$$
\begin{array}{ccc}
S_i & \xrightarrow{\quad I \quad} & I(S_i) \\
{\scriptstyle \star}\downarrow & & \downarrow{\scriptstyle \diamond} \\
\star_i S_i & \xrightarrow[\quad G \quad]{} & \diamond_i I(S_i).
\end{array}
$$

To this end, composition operators must be defined, as well as interesting local and global properties. Then, the difficult part is to find $\diamond$, that is, how composition propagates through properties. Moreover, it is not always possible to express the relationship between components and composed systems so easily: it is sometimes necessary to add some global information in addition to the local properties.

The method we propose and the techniques we introduce in this monograph are strongly influenced by program theory, where they prove most useful. In computing science, programs are defined as relations, and predicate-transformers express their semantics. Standard composition operators are introduced to build nondeterministic sequential programs, and their execution effect must be determined [91, 150, 93]. Parallel programs require further composition operators. There, the compositional analysis amounts to prove theorems extending specific properties from basic systems to composed ones [245, 61, 1, 67]. For instance, a program is represented by the set of its possible execution traces; composition is then also defined in terms of traces [216, 89].

Our work follows the same guidelines. We define systems in a relational setting, and we use set-transformers and execution traces to identify their dynamics. Composition operators serve to build structured systems from basic ones. Then, the compositional analysis of systems is carried out in order to study their dynamical complexity or to characterize their computational power: properties of composed systems are obtained by combination of individual properties of their components.

Abstraction. Although composition techniques may seem very powerful, a direct compositional analysis often remains intractable: a fully precise observation is not always realistic; a coarse-graining is often preferred since details can be omitted; etc.

The classical solution to this problem is abstraction, which consists in simplifying some inherently difficult features of a system, so that its analysis becomes possible. Under some assumptions, interesting qualitative properties are preserved, and conclusions at the abstract level are transferred back to the concrete level.

This technique is common and extensively used in both dynamical systems and program theory. For example, symbolic dynamics is based on the coarse-grained observation of evolutions of systems [139, 202]. Abstract interpretation [163], refinement [235], but also simulation and topological conjugacy [139, 326, 9], are all related to abstraction. In this monograph, we use abstract observation to simplify the study of dynamical properties, and simulation to compare the computational abilities of various classical models.

Case Studies: Complex Systems. To validate our approach, besides simple illustrations, we concentrate on three families of dynamical systems.

- Low-dimensional chaotic systems: Smale horseshoe map, logistic map, Cantor middle-thirds relation [88, 326]. These examples are prototypes of chaotic systems, and almost every textbook on chaos focuses on them. Analyzing them by composition is an unquestionable objective. Moreover, we must show this does help in understanding their behavior.

- Formal systems: paperfoldings [10, 13, 14, 83, 219, 268] model a physical realization of what happens in the previous family of systems: space is folded, which leads to chaos. Solving problems in formal systems is an interesting challenge, because they are fundamental tools of theoretical computing science, particularly in formal deductive processes.
- Spatiotemporally chaotic systems: cellular automata [330]. This last family is important for two reasons: first, such automata are typical models of distributed systems, where local interactions between processes lead to global effects; second, these systems show important complex phenomena called "spatiotemporal chaos" in the field of dynamical systems. Studying cellular automata can thus bring useful insights in computing science as well as in dynamical systems theory.

1.3 Overview of the Monograph

In this section, each chapter of the monograph is presented in a separate paragraph, and a brief description of its content is given. In Part One, Chap(s). 2–3 define the mathematical framework: iterated relations and composition. In Part Two, Chap(s). 4–5 define abstract complexity tools: abstraction, invariance and attraction. In Part Three, Chap(s). 6–9 present the main theoretical contributions as well as their applications to typical examples: the compositional analysis of dynamical and computational properties of systems. More precisely, Chap. 6 is devoted to dynamical complexity, Chap. 9 to computational power, and case studies in dynamical complexity are presented separately in Chap(s). 7–8. Finally, Chap. 10 concludes the study and proposes directions for future research. The global organization of the monograph is represented in Fig. 1.12, where arrows indicate prerequisite readings to understand technical points in a detailed way. Chapters are referred to by their numbers.

Chapter 2 gives a short introduction to *basic notions* in general topology and in the theory of dynamical systems. Of course, this chapter can be complemented by many excellent and more detailed textbooks [96, 66, 138, 88, 326, 9, 154].

Relational dynamical systems are defined, as well as their set-level deterministic, and point-level nondeterministic dynamics, respectively called set-transformers and trajectories. Fixpoint theorems and their respective assumptions are discussed, including a general scheme for transfinite iterations used in the following.

Chapter 3 introduces *composed dynamical systems*: they are obtained by composition of relational systems using specific operators defined in the chapter.

Compositional results are presented about the dynamics of composed systems. One-step evolutions and sets of trajectories are systematically analyzed for each operator. Algebraic properties of operators are examined, and their

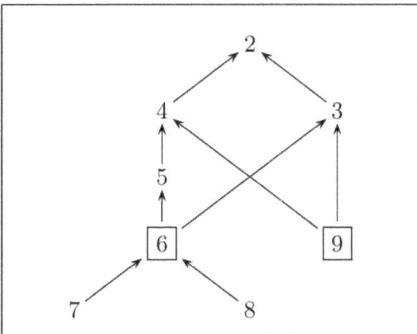

Fig. 1.12. Organization of the monograph. Arrows express requirements, and numbers represent chapters. Chap(s). 6 and 9 are the central parts of this monograph: dynamics and computations by abstraction and composition. Note Chap. 1 is the present introduction.

generalization leads to the definition of some relations as solutions of functional fixpoint equations.

Chapter 4 presents *abstract observations* of dynamics. First, the observation of a dynamical system induces a trace-based dynamics, where traces correspond to transition sequences. Then, the abstraction of observed trajectories by traces amounts to define an abstract system. This can be very useful as some qualitative dynamical properties, here invariance and reachability, are preserved by abstraction. They can thus be proved at the abstract level and still the conclusions remain valid at the concrete level under weak assumptions.

The coarse-grained aspect provided by traces, as well as these last conclusions, motivate their extensive use in the following.

Chapter 5 focuses on trace-based *dynamical properties* of systems, namely *invariance and attraction*, and relates them to dynamical complexity.

Finite (small) trajectories are not very interesting since they are often easily characterized; contrarily, complex behaviors arise from infinite trajectories. Invariants are sets of states containing arbitrarily long trajectories. Attraction is a complementary notion: it expresses relationships between initial and final states of (usually, infinite) trajectories.

Different types of invariants are defined and their structure is analyzed by means of two abstract observation-based properties: fullness (all abstract trajectories, i.e. traces, are realizable), and atomicity (the invariant of each trace contains at most one state). Fullness validates the use of a specific observation grain, whereas atomicity dually restricts the coarse-graining. Sufficient criteria are proposed to verify them, as their conjunction entails two features of chaos: sensitivity to initial conditions, and the existence of a trajectory between any pair of invariant states.

A taxonomy of attraction is given and Lyapunov-like sufficient criteria are then developed.

Chapter 6 analyzes *dynamical properties* of systems by *composition*. Each operator is systematically treated regarding invariance, attraction, and invariant structure. This chapter is thus central since it offers theoretical compositional results of composed dynamical systems. In particular, union deserves a special attention as it generates complex behaviors from elementary systems verifying a few assumptions: compatible dynamics on each independent subspace and symmetric attraction to different invariants generate fullness and atomicity. Actually, such complex behaviors can even emerge from very simple systems by structural composition if these assumptions are satisfied.

Chapter 7 analyzes *case studies* in the compositional analysis of dynamical properties of systems.

First, we rederive known results on the chaotic behavior of three systems by compositional analysis: Smale horseshoe map, Cantor relation, logistic map. The analysis of the last one is obtained from the Cantor relation analysis by successive transformations preserving qualitative compositional results.

Second, we concentrate on a specific family of formal systems: paperfoldings. By compositional analysis, we show that these systems have a Cantor-set invariant on which they behave chaotically.

All these examples are obtained by free product, union and sequential compositions.

Chapter 8 is entirely devoted to *cellular automata*, defined as connected products. First, we propose an attraction-based classification of behaviors. This leads to five classes that we structure in order to clearly isolate the class of most complex behaviors. A conjecture, obtained by simulation means, stated that complex cellular automata could be obtained by disjunction of shifting behaviors [49]. By compositional analysis and using additional complexity measures, we confirm the conjecture.

Chapter 9 examines *computational properties* of three classical systems by *compositional analysis*: Turing machines, cellular automata and continuous real functions are ordered in a strict hierarchy, which becomes an equivalence if we consider some limitations on systems (finite memory, finite computation time, approximations). These systems are embedded in a general model, based on the connected product, which is analyzed by composition regarding two types of properties. Extrinsic properties (simulation) only entail a weak hierarchy; intrinsic ones (continuity, shift-invariance, shift-vanishing, Lipschitz condition; (un)computability of initial conditions) strengthen the conclusion. This chapter extends previous results on cellular automata and computational models working on infinite objects [143, 273, 323].

Chapter 10 draws the *conclusions* of the study, summarizes the content and main contributions of the monograph, and proposes some research directions for the future.

2. Dynamics of Relations

Discrete-time dynamical systems are generally continuous functions defined on appropriate compact metric spaces. Their dynamics can be defined by successive iterations from any set of initial states, or as the set of all possible trajectories systems can follow.

Here, relations are considered instead of functions, for their ability of modeling nondeterminism, and the homogeneous mathematical framework they offer.

Indeed, nondeterminism is interesting to model many natural phenomena, mathematical concepts (difference or differential inclusions), or concrete problems in computing science (sequential computations and Dijkstra's guarded-command programs, parallel asynchronous systems, logic programs, and databases). Relations are closed under inversion, and allow backward evolutions symmetrically to forward evolutions.

Usual characteristics of dynamical systems entail stability, if ever, in at most ω steps. Working with relations weakens this property, and more than ω steps can be needed to stabilize their dynamics. Therefore, we introduce an iteration scheme based on transfinite ordinal numbers, which generalizes classical iteration schemes.

This chapter is essentially a summary of classical notions and results from relation algebra, program semantics, and dynamical systems theory: in §2.1, we introduce the notion of dynamical system and the dynamics based on infinite iterations of functions, and in §2.2, we give the relational version; in §2.3, we briefly summarize useful definitions and properties needed thereafter; in §2.4, we introduce transfinite iterations; finally, we close the chapter with a discussion in §2.5.

Before introducing this chapter, some useful notational conventions are first mentioned. They concern sequences of arbitrary spaces, or words of formal languages.

Notation 2.1 (Sequences, words). *Let X is be an arbitrary space (resp. an alphabet). Then $X^{\leq n}$ is the set of sequences (resp. words) of length smaller than n, X^n is the set of sequences of length n, X^* is the set of finite sequences of X (including the empty sequence ε), X^ω is the set of infinite sequences, $X^\infty = \Sigma^* \cup \Sigma^\omega$, $X^{n>\omega}$ is the set of sequences longer than ω, and $X^{\mathbb{O}} = X^* \cup X^\omega \cup X^{n>\omega}$. If s represents a sequence, $|s|$ denotes its length.*

F. Geurts: Abstract Compositional Analysis of Iterated Relations, LNCS 1426, pp. 21-52, 1998.
© Springer-Verlag Berlin Heidelberg 1998

Juxtaposition of symbols stands for concatenation; exponentiation stands for multiple concatenation. For any sequence s of length at least n, $s|_n$ represents its prefix of length n. For any bi-infinite $s \in X^{\mathbb{Z}}$, we denote the subsequence $s_0 s_1 s_2 \cdots$ by s^+ and $s_0 s_{-1} s_{-2} \cdots$ by s^-.

2.1 Functional Discrete-Time Dynamical Systems

Classically, a discrete-time dynamical system is a (generally, continuous) function f defined on a (generally, compact metric) space X (see e.g. [9]). The dynamical system imposes a temporal ordering on the underlying space. We interpret $f(x)$ as the point which immediately follows x in time.

In this first section, the dynamics of functional systems and related notions are just browsed; we give precise definitions in the more general setting of relations in §2.2.

The evolution of the system is based on successive iterations, starting from any initial state x_0 of X:

$$x_0 \xrightarrow{f} f(x_0) \xrightarrow{f} f(f(x_0)) \xrightarrow{f} \cdots$$

The iteration scheme is recursively defined as follows: $\forall x \in X, n \in \mathbb{N}$,

$$
\begin{aligned}
f^0(x) &= x \\
f^{n+1}(x) &= f(f^n(x)).
\end{aligned}
$$

This can be extended to sets of points easily, using a classical point-to-set lifting, or set-transformer: $\forall A \subseteq X$,

$$
\begin{aligned}
f &: \mathbb{P}(X) \mapsto \mathbb{P}(X) \\
\text{s.t.} \quad & f(A) = \cup_{a \in A} f(a)
\end{aligned}
$$

where $\mathbb{P}(X)$ denotes the *power set* of X, i.e. the set of all its subsets.

Notation 2.2. *For simplicity, we write $f(x)$ instead of $f(\{x\})$.*

Remark 2.3. Let us emphasize the (trivial) homomorphism exhibited by set-transformers. It expresses a point-to-set lifting, summarized by a commutative diagram:

$$
\begin{array}{ccc}
a & \xrightarrow{\ f\ } & f(a) \\
\cup \downarrow & & \downarrow \cup \\
\cup a & \xrightarrow[\ f\ ]{} & \cup f(a).
\end{array}
$$

Equivalently, the dynamics can also be expressed as the set of all possible evolutions that the system can follow from any initial state:

$$\theta(X, f) = \{s \in X^{\mathbb{N}} \mid \forall n, s_{n+1} = f(s_n)\}$$

or from initial states in a specific subset $A \subseteq X$:

$$\theta(A, f) = \{s \in X^{\mathbb{N}} \mid (s_0 \in A) \wedge (\forall n \in \mathbb{N}, s_{n+1} = f(s_n))\}.$$

Remark 2.4. – This definition is equivalent to the notion of continuous-time flow or discrete-time cascade [326].
– Each element $s \in \theta(A, f)$ is an ω-infinite trajectory.
– To avoid the problem of finite sequences for which the image of some s_n could be undefined, we assume that f is always defined. For example, in case f is undefined, we can add a special "undefined" symbol $\heartsuit$ to X such that $f(\heartsuit) = \heartsuit$.

Based on this, we see that initial conditions completely determine the dynamics of the system. Each state is univocally computed as the image of a previous state. Of course, this is only valid toward the future: computing backward histories is not always possible, because not all functions are invertible. For example, an injective function can be inverted and considered as a new function, but this is not a general case. In fact, the inverse of a noninvertible function is simply a relation. Iterating the system to the past gives rise to nondeterminism: several images can correspond to a single state.

Example 2.5. We consider the chaotic logistic map, $f(x) = 4x(1-x)$, defined on the unit interval $[0, 1]$ (see Fig. 2.1).

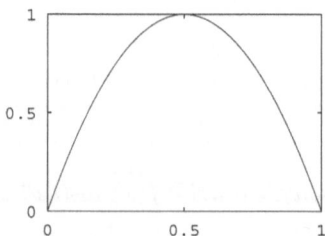

Fig. 2.1. Graph of $f(x) = 4x(1 - x)$

Among uncountably many others, four possible histories of the system, that is, elements of $\theta([0, 1], f)$, are represented in Fig. 2.2.

The inverse of f is not functional since f is 2-to-1. For example, the preimages of $\frac{8}{9}$ by f are $\frac{1}{3}$ and $\frac{2}{3}$. In other words, the image of $\frac{8}{9}$ by the inverse of f is equal to $\frac{1}{3}$ or $\frac{2}{3}$.

The aim of the next section is to extend the previous definitions to relations, in order to integrate nondeterminism elegantly.

2.2 Relational Dynamical Systems

In the previous section, we have shown the limitations of functional dynamical systems: they cannot be inverted, and nondeterministic choices cannot be treated systematically. Here, we extend the informal definitions given above to general relations, and we arrive at the notion of relational discrete-time dynamical system.

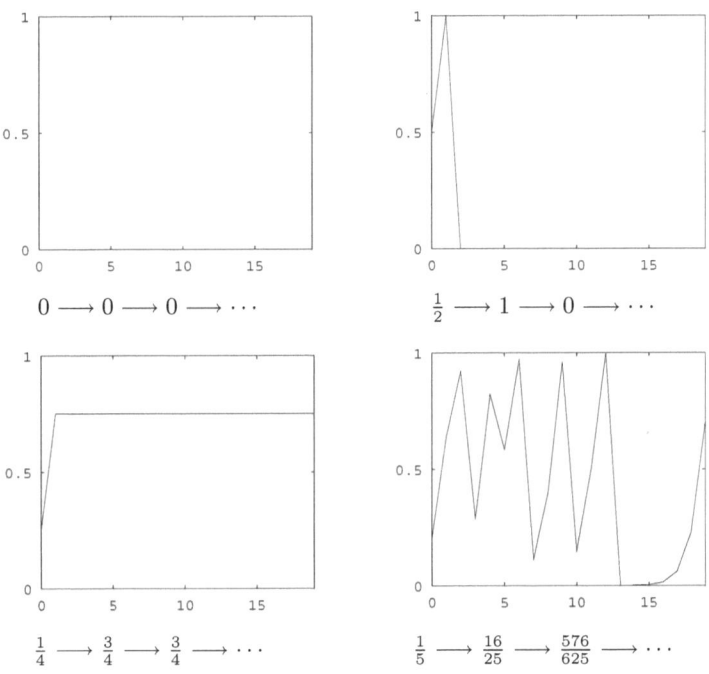

Fig. 2.2. Four possible evolutions of the logistic map $f(x) = 4x(1-x)$

Let f be a relation defined on a space X, i.e. $f \subseteq X \times X$. Each state of X can have zero, one or more images by f. Thus, *nondeterminism* is allowed, and the inverse of a relation being a relation, backward evolutions are allowed, too. More precisely, to fix the notation, the inverse of f is

$$f^{-1} = \{(y, x) \mid (x, y) \in f\}.$$

Usually, X and f are restricted to specific spaces and relations, in order to keep many results from elementary topology [9].

Definition 2.6 (Relational discrete-time dynamical system). *A relational discrete-time dynamical system (RDS) is a pair (X, f) where X is a compact metric space and f is a closed relation on X.*

Remark 2.7. – A *closed relation* on X is a closed subset of $X \times X$ (see also §2.3.2).
– Unless stated otherwise, all dynamical systems of this monograph will be based on that definition.
– A continuous function, regarded as a relation, is a closed relation. Closed relations naturally extend continuous functions. This assumption is discussed in [9] and in §2.3.2.

The iterative evolution from a state can be described from two viewpoints, emphasizing the intrinsic nondeterminism of relations at the point-level, or by means of multi-valued functions and set-transformers at the set-level.

2.2.1 Point-Level Nondeterministic Dynamics

Let (X, f) be a RDS, and $x \in X$. If there exists a y such that $(x, y) \in f$, then

$$x \xrightarrow{\ f\ } y$$

represents a possible iteration-step from x, and many other evolutions can coexist; f is regarded as a nondeterministic function from X to X.

The subsequent steps of the dynamics can be defined using this view, leading to the following definitions.

Definition 2.8 (Trajectory). An ω-infinite trajectory of a RDS (X, f) is a sequence $s \in X^{\mathbb{N}}$ such that, $\forall n \in \mathbb{N}, (s_n, s_{n+1}) \in f$.

Definition 2.9 (Nondeterministic forward dynamics). The nondeterministic forward dynamics of a RDS (X, f) from X is the set of all its trajectories starting from X:

$$\theta(X, f) = \{s \in X^{\mathbb{N}} \mid \forall n \in \mathbb{N}, (s_n, s_{n+1}) \in f\}.$$

From $A \subseteq X$, it is:
$$\theta(A, f) = A \times X^{\mathbb{N}} \cap \theta(X, f).$$

Now, this definition can also be extended to backward evolutions: it suffices to consider $\theta(A, f^{-1})$. Then, we define the complete (i.e. forward and backward) dynamics as follows.

Definition 2.10 (Nondeterministic dynamics). The nondeterministic dynamics of a RDS (X, f) from a set $A \subseteq X$ of initial conditions is

$$\Theta(A, f) = \{s \in X^{\mathbb{Z}} \mid (s_0 \in A) \wedge (s^+ \in \theta(A, f)) \wedge (s^- \in \theta(A, f^{-1}))\}.$$

Remark 2.11. Bi-infinite trajectories can be defined easily, using Def. 2.8.

2.2.2 Set-Level Deterministic Dynamics

Considering all possible images together leads to a deterministic view of the relation

$$x \xrightarrow{f} \{y \mid (x,y) \in f\}$$

and f is a deterministic multi-valued function, i.e. defined from X to $\mathbb{P}(X)$.

To compute subsequent steps of the evolution requires to apply f to any subset of X, because the image of any state is yet a set of states. This leads to the notion of set-transformer, which we have already used in the functional case.

Definition 2.12 (Set-transformer). *Let (X, f) be a RDS, then the set-transformer based on f is a function*

$$\begin{aligned} f \; &: \; \mathbb{P}(X) \mapsto \mathbb{P}(X) \\ \text{s.t.} \quad &f(A) = \cup_{a \in A} f(a). \end{aligned}$$

Remark 2.13. In the following, we do not make any distinction between "relations" and "set-transformers" because they are equivalent mathematical entities.

Set-transformers permit to define successive iterations from any subset A of X. The iteration scheme is recursively defined as follows: $\forall A \subseteq X, n \in \mathbb{N}$,

$$\begin{aligned} f^0(A) &= A \\ f^{n+1}(A) &= f(f^n(A)) \\ f^{-n}(A) &= (f^{-1})^n(A). \end{aligned}$$

Definition 2.14 (Deterministic forward dynamics). *The deterministic forward dynamics of a RDS (X, f) from a set $A \subseteq X$ of initial conditions is*

$$\xi(A, f) = (A_i)_i \text{ where } A_i = f^i(A), \forall i \in \mathbb{N}.$$

Again, the complete dynamics can be defined the same way.

Definition 2.15 (Deterministic dynamics). *The deterministic dynamics of a RDS (X, f) from a set $A \subseteq X$ of initial conditions is*

$$\Xi(A, f) = (A_i)_i \text{ where } A_i = f^i(A), \forall i \in \mathbb{Z}.$$

2.2.3 Comparison

In some sense, the set-level deterministic viewpoint is coarser than the point-level nondeterministic one because the former treats global evolutions together without relations between particular states, while the latter provides all possible specific evolutions.

To define the complete dynamics of a system, two choices are possible: either f, or the set of possible evolutions $\theta(X, f)$ must be given explicitly. Giving the sequence of successive iterates of X, viz. $\xi(X, f)$, is not enough, because one looses the relation between particular states and their image(s).

Example 2.16. Let us consider the logistic map on $[0,1]$ again or, more precisely, its inverse. We obtain the equations of its two branches by solving $f(x) = y$. We get $f^{-1}(y) = \frac{1}{2}(1 \pm (1-y)^{\frac{1}{2}})$; all points but 1 have two images by f^{-1} which is clearly relational (see Fig. 2.3). As we already know,

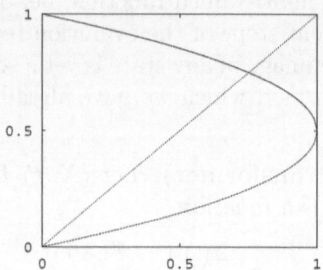

Fig. 2.3. Inverse logistic map

$f^{-1}(\frac{8}{9}) = \{\frac{1}{3}, \frac{2}{3}\}$. The images of these points are $f^{-1}(\frac{1}{3}) = \{\frac{1}{2}+(\frac{1}{6})^{\frac{1}{2}}, \frac{1}{2}-(\frac{1}{6})^{\frac{1}{2}}\}$ and $f^{-1}(\frac{2}{3}) = \{\frac{1}{2}+(\frac{1}{12})^{\frac{1}{2}}, \frac{1}{2}-(\frac{1}{12})^{\frac{1}{2}}\}$. And so on, and so forth.

Giving the dynamics in terms of possible evolutions goes as follows:

$$
\theta(\{\tfrac{8}{9}\}, f^{-1}) \;=\; \{\tfrac{8}{9} \longrightarrow \tfrac{1}{3} \longrightarrow \tfrac{1}{2} + (\tfrac{1}{6})^{\frac{1}{2}} \longrightarrow \cdots,
$$
$$
\tfrac{8}{9} \longrightarrow \tfrac{1}{3} \longrightarrow \tfrac{1}{2} - (\tfrac{1}{6})^{\frac{1}{2}} \longrightarrow \cdots,
$$
$$
\tfrac{8}{9} \longrightarrow \tfrac{2}{3} \longrightarrow \tfrac{1}{2} + (\tfrac{1}{12})^{\frac{1}{2}} \longrightarrow \cdots,
$$
$$
\tfrac{8}{9} \longrightarrow \tfrac{2}{3} \longrightarrow \tfrac{1}{2} - (\tfrac{1}{12})^{\frac{1}{2}} \longrightarrow \cdots,
$$
$$
\cdots\}.
$$

Determinizing the system by considering its global evolution from the initial condition gives:

$$
\{\tfrac{8}{9}\} \longrightarrow \{\tfrac{1}{3}, \tfrac{2}{3}\} \longrightarrow \{\tfrac{1}{2} + (\tfrac{1}{6})^{\frac{1}{2}}, \tfrac{1}{2} - (\tfrac{1}{6})^{\frac{1}{2}}, \tfrac{1}{2} + (\tfrac{1}{12})^{\frac{1}{2}}, \tfrac{1}{2} - (\tfrac{1}{12})^{\frac{1}{2}}\} \longrightarrow \cdots
$$

It is clear that the set of possible evolutions gives as much information as the relation itself, whereas the evolution of sets hides some important relationships between states. For example, the last expression does not show that it is not possible to go from $\frac{2}{3}$ to $\frac{1}{2} + (\frac{1}{6})^{\frac{1}{2}}$.

Before concluding this section, let us show that to know the relation

$$
f \subseteq X \times X
$$

and to know its dynamics

$$\theta(X, f) \in \mathbb{P}(X^{\mathbb{N}})$$

are equivalent, and that they both permit to define the sequence of successive iterations from the whole space X

$$\xi(X, f) \in (\mathbb{P}(X))^{\mathbb{N}}.$$

Proposition 2.17. *Let (X, f) be a RDS, then*

– *the knowledge of f and $\theta(X, f)$ are equivalent;*
– *the knowledge of $\theta(X, f)$ permits to define $\xi(X, f)$.*

Proof. – Given f, $\theta(X, f)$ is obtained by Def. 2.9.
– Given $\theta(X, f)$, f is defined by

$$f = \{(s_0, s_1) \mid \exists s \in \theta(X, f)\}.$$

– Given $\theta(X, f)$, $\xi(X, f)$ is defined by

$$\xi(X, f) = (X_i)_i \text{ where } X_i = \{s_i \mid \exists s \in \theta(X, f)\}, \forall i \in \mathbb{N}.$$

2.3 Preliminary Definitions and Properties

This section recalls some standard definitions and properties of relation algebra, topology, and calculus, in order to keep this monograph reasonably self-contained. Our sources are [96] for topology, [276] for relations, [78] for lattice theory and [91, 93] for predicate-transformer semantics of programs.

2.3.1 Basic Definitions About Relations

In general, a relation is a subset of a Cartesian product of spaces. According to [277], a *homogeneous relation* on a space X is a subset of $X \times X$, and a *heterogeneous relation* between X and Y is a subset of $X \times Y$. Let us denote the sets of homogeneous relations on X as

$$\mathcal{R}(X) = \mathbb{P}(X \times X)$$

and the set of heterogeneous relations between X and Y as

$$\mathcal{R}(X, Y) = \mathbb{P}(X \times Y).$$

The domain and range of a relation f are defined as follows.

Definition 2.18 (Domain, range). *The domain and range of a relation f are*

$$
\begin{aligned}
Dom(f) &= \{x \mid \exists y, (x,y) \in f\} \\
Rg(f) &= \{y \mid \exists x, (x,y) \in f\}.
\end{aligned}
$$

A notion of recurrence is obtained as follows.

Definition 2.19 (Fixpoint, periodic point). *Let f be a relation on X, $x \in X$ and $n \in \mathbb{N}\backslash\{0\}$. Then, x is a fixpoint of f iff $x \in f(x)$. It is (strictly) n-periodic iff $x \in f^n(x)$ (and $\forall m \in \{1, \cdots, n-1\}, x \notin f^m(x)$).*

The simplest relations one can think of are the following ones.

Definition 2.20 (Empty, universal, identity relations). *The empty, universal and identity relations are given by:*

$$
\begin{aligned}
\mathcal{E} &= \emptyset \\
\mathcal{U}_{X,Y} &= X \times Y \\
\mathcal{I}_X &= \{(x,x) \mid x \in X\}.
\end{aligned}
$$

In the following, indices will be removed when clear from the context.

The following proposition expresses that no image other than the empty set can come from an application of the empty relation. The same holds when any relation is applied to the empty set. This is the reason it is called "excluded miracle".

Proposition 2.21 (Excluded miracle). *Let f be a relation on X, and $A \subseteq X$; then*

$$
\begin{aligned}
\mathcal{E}(A) &= \emptyset \\
f(\emptyset) &= \emptyset.
\end{aligned}
$$

Proof. The first part is trivial:

$$
\mathcal{E}(u) = \{v \mid (u,v) \in \mathcal{E}\} = \emptyset.
$$

The second is also direct:

$$
f(\emptyset) = \cup_{u \in \emptyset} f(u) = \emptyset.
$$

We have also very similar properties concerning $\mathcal{U}$ and $\mathcal{I}$, stated without proof.

Proposition 2.22. *For all nonempty $A \subseteq X$,*

$$
\begin{aligned}
\mathcal{U}_{X,Y}(A) &= Y \\
\mathcal{I}_X(A) &= A.
\end{aligned}
$$

As in a functional framework, relations can be characterized regarding the number of images or preimages they have.

Definition 2.23 (Types of relations). *A relation $f \in \mathcal{R}(X)$ is called total iff*

$$\forall u, \#f(u) \geq 1;$$

simple (i.e. functional) iff

$$\forall u, \#f(u) \leq 1;$$

finite iff

$$\forall u, \#f(u) \in \mathbb{N};$$

surjective iff

$$\forall v, \#f^{-1}(v) \geq 1;$$

injective iff

$$\forall v, \#f^{-1}(v) \leq 1;$$

inverse finite iff

$$\forall v, \#f^{-1}(v) \in \mathbb{N};$$

constant iff

$$\#Rg(f) = 1;$$

inverse constant iff

$$\#Dom(f) = 1.$$

Later on, we will need the concept of projection. Observe that it can also be applied to sets that are not considered as relations.

Definition 2.24 (Projection). *If $E = \times_{i \in J} X_i$ is a Cartesian product, $u \in E$ is an element of E, $r \subseteq E$ is a subset of E, $I \subseteq J$ is a subset of indices, and $R \in \mathcal{R}(A, J)$ is a relation of indices ($\forall i \in A, R(i) = \{j \mid (i,j) \in R\} \subseteq J$), the projections of u and r on the indices I and R are given by*

$$
\begin{aligned}
\Pi_I(u) &= (u_i)_{i \in I} \\
\Pi_R(u) &= (\Pi_{R(i)}(u))_{i \in A} \\
\Pi_I(r) &= \cup_{u \in r} \Pi_I(u) \\
\Pi_R(r) &= \cup_{u \in r} \Pi_R(u).
\end{aligned}
$$

Elements of R are seen as vectors of E, the projector deletes all components indexed in $J \setminus I$ and only retains components indexed in I. This definition of course applies to relations, too.

Example 2.25. If r is a ternary relation on X, i.e. $r \subseteq X \times X \times X$, its projection on positions 1 and 3 is:

$$\Pi_{1,3}(r) = \{(u, w) \mid \exists v, (u, v, w) \in r\}.$$

2.3.2 Notions from Topology

Definition 2.26 (Topology). *A topology on a space X is a family T of subsets of X, called* open sets, *such that*

- *T is closed under union and finite intersection;*
- *$\emptyset$ and X are in T.*

Example 2.27. Two extreme examples of topological spaces are the *indiscrete topology* on X, i.e. $(X, \{\emptyset, X\})$, and the *discrete topology*, i.e. $(X, \mathbb{P}(X))$.

Let us now enumerate a list of useful definitions and properties.

Limits.

- A *neighborhood* N of a state $x \in X$ is an open set containing x; we denote it by N_x.
- A sequence $(y_i)_i$ of A *converges* to y iff every neighborhood U of y is such that $\exists k, \forall n \geq k, y_n \in U$.
- A sequence $(y_i)_i$ of A *accumulates* at y iff every neighborhood U of y is such that $\forall k, \exists n \geq k, y_n \in U$.
- An *isolated point* $a \in A$ is such that no sequence of distinct elements of A converges to a.

Sets.

- A *closed set* is the complement of an open set in the same topology.
- A set A is closed iff the limit of every convergent sequence of A belongs to A.
- Arbitrary intersections and finite unions of closed sets are closed.
- The *closure* of a set A is the union of A and the limit of all sequences of A. We denote it by $\overline{A}$.
- A set A is a *dense set* in B iff its closure is equal to B, i.e. $\overline{A} = B$.
- Any Cartesian product of closed sets is closed.
- A *perfect set* is a closed set that contains no isolated point.
- A set $A \subseteq X$ is a *connected set* if it is not the union of two nonempty disjoint open sets.
- The *component* of a point $x \in X$ is the union of all connected sets containing x.
- A *totally disconnected set* $A \subseteq X$ is such that $\forall x \in A$, the component of x is $\{x\}$.
- A *Hausdorff space* is such that each two distinct points have nonintersecting neighborhoods. Any finite power of the Euclidean space is Hausdorff. Discrete spaces are Hausdorff. Nontrivial indiscrete spaces are not Hausdorff.

Functions.

- A function defined on X is a *continuous function* iff the inverse image of any open set is an open set.
- A *homeomorphism* is a bijective continuous function such that its inverse is also continuous.

Compactness.

- A Hausdorff space is a *compact space* if each covering by open sets has a finite subcovering.
- A set A is compact iff every sequence in A has a convergent subsequence whose limit is in A, iff every sequence accumulates in A.
- A discrete space is compact iff it is finite.
- In any space, all finite subsets, and the empty set, are compact sets.
- A finite union of compact sets is a compact set.
- A subset of a compact space is compact iff it is closed.
- Any compact set of a metric space is closed and bounded.
- The continuous image of a compact set is compact.
- Any Cartesian product of compact spaces is compact (*Tychonoff lemma*).
- Any compact metric space is also complete, that is, any Cauchy sequence converges.

A very important type of set we will use is based on three definitions given above [96].

Definition 2.28 (Cantor set). *A Cantor set is a closed, totally disconnected, perfect set.*

Example 2.29 (Cantor middle-thirds set). We consider $[0, 1]$. Let us remove the open interval $(\frac{1}{3}, \frac{2}{3})$. From the remaining intervals, we remove the middle thirds $(\frac{1}{9}, \frac{2}{9})$ and $(\frac{7}{9}, \frac{8}{9})$, and repeat this process ad infinitum. The result is the famous Cantor middle-thirds set (see Fig. 2.4).

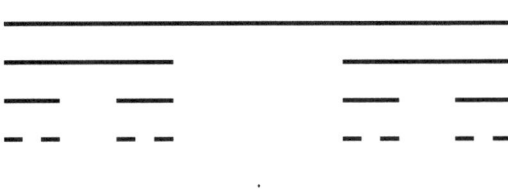

Fig. 2.4. Iterative construction of Cantor's middle-thirds set: recursive elimination of middle thirds intervals

Finally, relations are assumed to be closed subsets of the Cartesian space where they are defined. The following proposition is stated without proof.

Proposition 2.30. *Let* (X, f) *be a RDS and* A *be a closed subset of* X. *Then,* $f(A)$ *and* f^{-1} *are closed.*

This closure assumption extends functional continuity in the following sense. We know from topology that a function is continuous if and only if the inverse image of any open set is an open set. For relations the following weaker property holds [9].

Proposition 2.31. *Let* $f \in \mathcal{R}(X, Y)$ *be a closed relation, and* $U \subseteq Y$ *be an open set, then* $\{x \mid f(x) \subseteq U\} \subseteq X$ *is an open set.*

Proof. If f is a closed relation, and A is a closed subset of X, then f^{-1} is closed and $f(A)$ is closed, too. Thus, if B is closed, then $f^{-1}(B)$ is closed. The last set can be rewritten as $\{x \mid f(x) \cap B \neq \emptyset\}$. Since U is open, it complement $Y \backslash U$ is closed, and $f^{-1}(Y \backslash U)$ is closed, too.

Thanks to Def. 2.6, we have several interesting properties, summarized in the following proposition.

Proposition 2.32. *Let* (X, f) *be a RDS. Then,* $Dom(f)$ *and* $Rg(f)$ *are compact subsets of* X.

Proof. The space X is compact, and f is closed. Thus, $Dom(f)$ and $Rg(f)$ are closed, hence compact subsets of X.

The dynamics of a system is closed whenever the underlying relation is closed, which is the case in all RDS.

Proposition 2.33. *If* (X, f) *is a RDS, then* $\theta(X, f)$ *is closed in* $X^{\mathbb{N}}$.

Proof. By Def. 2.6, f must be closed. Hence, if we have $N_x \times N_y \cap f \neq \emptyset$ for all $x, y \in X$, and every open neighborhood N_x of x and N_y of y, then $(x, y) \in f$.

Let s belong to the adherence of $\theta(X, f)$, and N_i, N_{i+1} be two open neighborhoods of s_i and s_{i+1} respectively. For all $j \neq i, i+1$, we choose an open neighborhood N_j of s_j. We have thus $\times_i N_i \cap \theta(X, f) \neq \emptyset$, which implies $N_i \times N_{i+1} \cap f \neq \emptyset$. Thus, (s_i, s_{i+1}) is adherent to f, whence it belongs to f, from which we conclude that $s \in \theta(X, f)$.

This means that $\theta(X, f)$ is closed.

2.3.3 Monotonicity and General Junctivity Properties

Considering set-transformers, monotonicity can be seen in two ways.

Proposition 2.34 (Monotonicity). *Let* f *and* g *be two relations on* X, *and* $A \subseteq X$, *then*

$$f \subseteq g \Rightarrow f(A) \subseteq g(A).$$

Proof.

$$x \in f(A)$$
$$\equiv \quad \exists u \in A, x \in f(u)$$
$$\equiv \quad \exists u \in A, x \in \{v \mid (u, v) \in f\}$$
$$\because \quad \text{Hyp. } f \subseteq g$$
$$\Rightarrow \quad \exists u \in A, x \in \{v \mid (u, v) \in g\}$$
$$\equiv \quad x \in g(A).$$

Proposition 2.35 (Monotonicity). *Let f be a relation on X, and $A, B \subseteq X$, then*

$$A \subseteq B \Rightarrow f(A) \subseteq f(B).$$

Proof.

$$x \in f(A)$$
$$\equiv \quad \exists u \in A, x \in f(u)$$
$$\because \quad \text{Hyp. } A \subseteq B$$
$$\Rightarrow \quad \exists u \in B, x \in f(u)$$
$$\equiv \quad x \in f(B).$$

The following trivial result involves monotonic relations.

Proposition 2.36. *Let f be a relation, and $(X_i)_i$ be any sequence of subsets of X. Then*

$$f(\cap_i X_i) \quad \subseteq \quad \cap_i f(X_i)$$
$$f(\cup_i X_i) \quad \supseteq \quad \cup_i f(X_i).$$

Proof. We have

$$\forall i, \cap_i X_i \subseteq X_i,$$

and monotonicity of f (Prop. 2.35) gives

$$\forall i, f(\cap_i X_i) \subseteq f(X_i).$$

This entails $f(\cap_i X_i) \subseteq \cap_i f(X_i)$.
By monotonicity, $\forall i$,

$$X_i \subseteq \cup_i X_i \Rightarrow f(X_i) \subseteq f(\cup_i X_i).$$

Hence, $\cup_i f(X_i) \subseteq f(\cup_i X_i)$.

Stronger properties are interesting, where inclusions are replaced by equalities. Intersection is equivalent to conjunction, and union is equivalent to disjunction, whence the generic term "junctivity". Such junctivity properties are useful when using various fixpoint theorems. All of them are further discussed in [284, 93].

Definition 2.37 (Junctivity types). *Let f be a relation on X, and $V \subseteq \mathbb{P}(X)$. Then, f is conjunctive over V iff*

$$f(\cap_{A \in V} A) = \cap_{A \in V} f(A)$$

and disjunctive over V iff

$$f(\cup_{A \in V} A) = \cup_{A \in V} f(A).$$

It is universally junctive if the property holds for every V, positively junctive for any nonempty V, denumerably junctive for any nonempty countable V, finitely junctive for any nonempty finite V, and-continuous/or-continuous for any nonempty linear V (its elements can be arranged in a monotonic sequence), monotonic for any nonempty linear finite V.

These different forms of junctivity are related to each other by the following proposition [93].

Proposition 2.38. *Universal conjunctivity $\Rightarrow$ positive conjunctivity.*
Positive conjunctivity $\Rightarrow$ denumerable conjunctivity.
Denumerable conjunctivity $\equiv$ finite conjunctivity $\wedge$ and-continuity.
Finite conjunctivity $\vee$ and-continuity $\Rightarrow$ monotonicity.
Finite conjunctivity $\wedge$ or-continuity $\Rightarrow$ and-continuity.

In case of our set-transformers, we get or-continuity for free.

Proposition 2.39 (Universal disjunctivity, or-continuity). *Let (X, f) be a RDS, then the corresponding set-transformer is or-continuous, and even universally disjunctive.*

Proof. Given Def. 2.12, we have $f(\cup_i A_i) = \cup_i f(A_i)$.

Contrarily, we get and-continuity if and only if the nondeterminism of the relation is bounded, that is, each relation is finite. In [284], it is proved that bounded nondeterminism is a necessary and sufficient condition for and-continuity. The author works with predicate-transformers that match our definitions of set-transformers. Stated in our framework, this gives the following proposition [284, Prop. 8].

Proposition 2.40 (And-continuity). *The set-transformer based on a relation f defined on a space X is and-continuous iff f is inverse finite, and its inverse f^{-1} is and-continuous iff f is finite.*

Proof. Let us prove that f^{-1} is and-continuous iff f is finite.

$\Leftarrow$Let $(X_i)_i$ be a decreasing sequence of sets, i.e. $\forall i, X_{i+1} \subseteq X_i$. As f and, thus, f^{-1}, are monotonic, we have

$$\cap_i X_i \subseteq X_i$$
$$\Rightarrow \quad f^{-1}(\cap_i X_i) \subseteq f^{-1}(X_i)$$
$$\Rightarrow \quad f^{-1}(\cap_i X_i) \subseteq \cap_i f^{-1}(X_i).$$

To prove the reverse implication, the following equivalences hold:

$$x \in \cap_i f^{-1}(X_i)$$
$$\equiv \quad \forall i, x \in f^{-1}(X_i)$$
$$\equiv \quad \forall i, \exists y_i \in X_i, (y_i, x) \in f^{-1}$$
$$\equiv \quad \forall i, \exists y_i \in \cap_{j \leq i} X_j, (y_i, x) \in f^{-1}.$$

As f is finite, there is only a finite number of such y_i. Since $(X_i)_i$ is decreasing, one of these y_i belongs to all X_i. Thus,

$$\exists y_j \in \cap_i X_i, (y_j, x) \in f^{-1}$$
$$\equiv \quad x \in f^{-1}(\cap_i X_i).$$

$\Rightarrow$Let us now suppose that f is not finite. Let x be such that $\exists (y_i)_i$ an infinite sequence of distinct states of $f^{-1}(x)$. For each i, we define $P_i = \{y_i, y_{i+1}, y_{i+2}, \cdots\}$. By construction, $(P_i)_i$ is a decreasing sequence, whose intersection is empty. Thus $f^{-1}(\cap_i P_i) = f^{-1}(\emptyset) = \emptyset$. On the other hand, $\forall i, x \in f^{-1}(P_i)$, and $x \in \cap_i f^{-1}(P_i)$, which entails a contradiction.

Remark 2.41. This slightly differs from [92], where the or-continuity of the predicate-transformer wp is also restricted to bounded nondeterminism. This difference is due to the fact that, stated in our framework, R being a relation on X and $P \subseteq X$, $wp \cdot R \cdot P$ is not equivalent to $R^{-1}(P)$ but

$$wp \cdot R \cdot P = R^{-1}(P) \cap (X \backslash R^{-1}(X \backslash P)).$$

The set-difference modifies the result because it is no more or-continuous, since it is even not monotonic but anti-monotonic. Of course, and-continuity suffers from the same drawback.

Due to the type of space and relation we use in our framework, namely closed relations on compact metric spaces, and-continuity can be obtained without bounded nondeterminism. The trade-off comes from the new definition we give below.

Definition 2.42 (And-continuity*). Let (X, f) be a RDS. The set-transformer f is and-continuous* iff for any decreasing sequence of closed subsets of X, $(X_i)_i$ such that $\forall i, X_{i+1} \subseteq X_i$, we have

$$f(\cap_i X_i) = \cap_i f(x_i).$$

The only differences with and-continuity are the requirements of (X, f) being a RDS and the X_i's being closed sets. Based on this, we have the following proposition (see also [9, p. 9]).

Proposition 2.43 (And-continuity*). *Let* (X, f) *be a RDS, then* f *is and-continuous*.*

Proof. $\supseteq$ By Prop. 2.36, we have $\cap_i f(X_i) \supseteq f(\cap_i X_i)$.
$\subseteq$ Let $(X_i)_i$ be a decreasing sequence of sets. Let y be in $\cap_i X_i$: $\forall i, y \in f(X_i)$.
Thus, $\forall i, \exists x_i \in X_i \cap Dom(f), (x_i, y) \in f$.
Let us define

$$Y_i = \{x \in X_i \cap Dom(f) \mid (x, y) \in f\} = X_i \cap Dom(f) \cap f^{-1}(y).$$

Each Y_i is nonempty, i.e. contains at least one element x_i, and contained in X_i. Since $X_{i+1} \subseteq X_i$ for all i, we have also $Y_{i+1} \subseteq Y_i$. Let us prove that $\cap_i Y_i \neq \emptyset$.

– The space X is compact, and $Dom(f)$ is also compact since f is a closed relation (Prop. 2.32). If we choose one $x_i \in Y_i$ for each i, we construct a sequence in $Dom(f)$ which has an accumulation point x:

$$\forall N_x, \forall k, \exists n \geq k, x_n \in N_x.$$

– This accumulation point $x \in \cap_i Y_i$. If this is not the case, $\exists l, x \notin Y_l$. Thus, $\forall k > l, x \notin Y_k$ since $Y_k \subseteq Y_l$. Since X_l is closed, so is Y_l because $f^{-1}(y)$ is closed. Thus, $\exists N_x, N_x \cap Y_l = \emptyset$, and $\forall k \geq l, N_x \cap Y_k = \emptyset$. This contradicts the accumulation at x.

Thus, we have $x \in Dom(f)$ and $\cap_i Y_i \neq \emptyset$ since $x \in \cap_i Y_i$. Hence $y \in f(\cap_i X_i)$, and

$$\cap_i f(X_i) \subseteq f(\cap_i X_i).$$

In [284, 92], and-continuity was proved equivalent to bounded nondeterminism. In compact metric spaces, the stronger "and-continuity*" property can be obtained without satisfying this assumption, and lattice fixpoint theorems remain valid.

2.3.4 Fixpoint Theorems

In the rest of this monograph, convergence will be required for several aspects of the dynamics of systems (invariance, attraction). To justify this convergence, fixpoint theorems are always used. We present here a very important fixpoint theorem known as *Knaster-Tarski's theorem*. Its interest is to require few assumptions, viz. essentially a lattice structure. Then, we give a constructive version of the theorem. (We refer to [306, 197] for different historical versions of this "folk" theorem.)

We work with set-transformers acting on sets of a space X, i.e. we work with elements of $\mathbb{P}(X)$. This power set happens to be a *complete lattice*, which has an order defined by the set inclusion $\subseteq$; a bottom element, the empty set $\emptyset$; a top element, the space itself X; for each set $A \subseteq \mathbb{P}(X)$ of sets, a greatest lower bound given by the intersection $\cap_{B \in A} B$, and a least upper bound given by the union $\cup_{B\in A}B$. We have a complete lattice $\mathbb{P}(X)(\subseteq, \emptyset, X, \cap, \cup)$. More generally, we will denote such a *lattice* by $\mathbb{L}(\leq, \bot, \top, \sqcap, \sqcup)$ with *order relation* $\leq$, *bottom element* $\bot$, *top element* $\top$, *greatest lower bound* $\sqcap$ and *least upper bound* $\sqcup$.

Theorem 2.44 (Lattice-theoretical fixpoint theorem). *Let $\mathbb{L}$ be a complete lattice, f be a monotonic function on $\mathbb{L}$, and P be the set of fixpoints of f, then P is not empty and it is also a complete lattice. In particular, the least upper (resp. greatest lower) bound of P belongs to P and is equal to the least upper (resp. greatest lower) bound of the set $\{A \mid A \leq f(A)\}$ (resp. $\{A \mid f(A) \leq A\}$).*

Monotonicity is thus an important property (see §2.3.3). Adding the hypothesis of or-continuity (resp. and-continuity; again, see §2.3.3), the least (resp. greatest) fixpoints can be reached by successive iterations, starting from the bottom (resp. top) element of the lattice.

Definition 2.45 (Decreasing sequence, increasing sequence). *A sequence of a complete lattice $\mathbb{L}$, $(X_i)_{i\in\mathbb{N}}$, is decreasing iff $\forall i, X_{i+1} \leq X_i$. Is is increasing iff $\forall i, X_i \leq X_{i+1}$.*

Theorem 2.46 (Constructive lattice fixpoint theorem). *If $\mathbb{L}$ is a complete lattice, f is or-continuous on $\mathbb{L}$, and P is the set of its fixpoints, then the least fixpoint can be reached by successive increasing iterations from the bottom element:*

$$\mathbb{S}f = \sqcap P = \sqcup_{i<\omega} f^i(\bot).$$

If, f is and-continuous, the greatest fixpoint can be obtained as limit of successive decreasing iterations from the top element:

$$\mathbb{G}f = \sqcup P = \sqcap_{i<\omega} f^i(\top).$$

Remark 2.47. – The first sequence of iterations is increasing since f is monotonic and $\bot \leq f(\bot)$. In the same way, the second sequence is decreasing since $f(\top) \leq \top$.

– The results of these theorems can be generalized to any complete partial order. Of course, in this case, only one direction of iteration is possible, upward or downward.

– The results also apply to $\mathbb{K}(X)$, i.e. the set of nonempty compact subsets of X, instead of $\mathbb{P}(X)$. Moreover, whereas in a power set $\mathbb{P}(X)$ the bottom element is the emptyset, it is not the case in $\mathbb{K}(X)$.

– The theorem still holds when and-continuity is replaced by and-continuity*, $\mathbb{L}$ is equal to $\mathbb{P}(X)$ or to $\mathbb{K}(X)$, and X is a compact metric space. In case of RDS, all these assumptions are verified (see Def. 2.42 and Prop. 2.43).

2.3.5 Elementary Properties

The following propositions will be used later.

Proposition 2.48. *If a relation f is injective, we have $\forall u \neq v, \forall A, B \subseteq X$,*

$$f(u) \cap f(v) = \emptyset$$
$$f(A \cap B) = f(A) \cap f(B).$$

Proof. If the first line is false, this means that the intersection is not empty. Thus, there exists y such that $y \in f(u)$ and $y \in f(v)$, which means $\{u, v\} \subseteq \{x \mid (x, y) \in f\}$ and contradicts the hypothesis.

We have one inclusion by monotonicity: $f(A \cap B) \subseteq f(A) \cap f(B)$ (Prop. 2.36). The other ones goes as follows. Let y be in $f(A) \cap f(B)$. This means $y \in f(A)$ and $\exists u \in A, (u, y) \in f$. The same holds for B: $\exists v \in B, (v, y) \in f$. Since f is injective, u must be equal to v. Thus, $\exists u \in A \cap B, (u, y) \in f$ and $y \in f(A \cap B)$. $\blacksquare$

Inversion is not always strict. In fact, we have only a semi-inversion property.

Proposition 2.49 (Semi-inversion). *Let f be a relation on X, and $A \subseteq X$, then*

$$A \subseteq f^{-1}(f(A)).$$

Instead of proving this trivial property, let us show that the implication can be strict.

Example 2.50. Consider for instance a relation $ex \subseteq \mathbb{R} \times \mathbb{R}$ containing two vertical segments of the plane $x - y$: $ex = \{1\} \times [2, 5] \cup \{2\} \times [2, 5]$. (see Fig. 2.5). From 1, we have $ex(\{1\}) = [2, 5]$ and $ex^{-1}([2, 5]) = \{1, 2\}$. From a strict subset

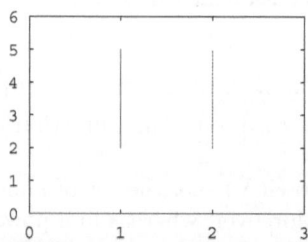

Fig. 2.5. Graph of $ex = \{1\} \times [2, 5] \cup \{2\} \times [2, 5]$

of $[2, 5]$, $[3, 4]$, we have $ex^{-1}([3, 4]) = \{1, 2\}$ and $ex(\{1, 2\}) = [2, 5]$.

In Prop. 2.49, "$\subseteq$" can be replaced by "$=$" under a stronger assumption on the relation, that has to be injective and total. We state it without proof.

Proposition 2.51. *Let f be a total injective relation on X, g be a surjective simple relation (i.e. function), and $A \subseteq X$. Then,*

$$A = f^{-1}(f(A)) \text{ and } A = g(g^{-1}(A)).$$

2.3.6 Metric Properties

Useful properties can be defined for some relations, according to the way they expand or contract their domain. To this end, we have to precisely define the distance for each space we work in, and the properties of these metrics.

First, let us recall two distances we will intensively use in the following.

Definition 2.52 (Euclidean distance). *In the set of real number $\mathbb{R}$, the Euclidean distance between two points x and y is*

$$d_e(x, y) = |x - y|.$$

Definition 2.53 (Astronomer's metric). *In any set of infinite sequences $X^{\mathbb{N}}$, the distance between two elements $x = (x_i)_i$ and $y = (y_i)i$ is*

$$d_a(x, y) = 2^{-\inf\{i | x_i \neq y_i\}}.$$

Remark 2.54. This last distance is called "astronomer's metric" because the further differences happen from the origin, the smaller the weights are in the contribution to the distance.

Definition 2.55 (Diameter). *Let (X, d) be a metric space. Then, for any subset $A \subseteq X$, its diameter is*

$$diam(A) = \sup_{x,y \in A} d(x, y).$$

In general, we use a RDS, which involves a compact metric space (X, d), where d is the metric defined on X. Relations are regarded as multi-valued functions from X to $\mathbb{P}(X)$. This requires a metric on $\mathbb{P}(X)$. We consider the standard Hausdorff metric.

Definition 2.56 (Hausdorff metric). *Let (X, d) be a metric space. The Hausdorff metric h on $\mathbb{P}(X)$ is given as follows: $\forall A, B \in \mathbb{P}(X)$,*

$$h(A, B) = \max\{h'(A, B), h'(B, A)\}$$

where

$$h'(A, B) = \sup_{x \in A} h''(x, B)$$

$$h''(x, B) = \inf_{y \in B} d(x, y).$$

The following result is proved in [159] and [28, Theorem 2.7.1, p. 37].

Proposition 2.57. *If* (X, d) *is a complete (resp. compact) metric space, then* $(\mathbb{K}(X), h)$ *is a complete (resp. compact) metric space, too.*

This means that working with nonempty compact sets of X leads to the same topological and metric properties as working with states of X. Since closed relations preserve compactness, we can restrict ourselves to compact subsets.

The following properties will be useful [28, 325].

Proposition 2.58. *Let* A, B, C *be subsets of* X, *then*

$$B \subseteq C \Rightarrow h'(A, C) \leq h'(A, B).$$

Proof. We know that $h'(A, C) = \sup_{x \in A} \inf_{y \in C} d(x, y)$. Since $B \subseteq C$, for each x, we have $\inf_{y \in C} d(x, y) \leq \inf_{y \in B} d(x, y)$. By monotonicity of sup, we get the result.

Proposition 2.59. *Let* A, B, C *be subsets of* X, *then*

$$h'(A \cup B, C) = \sup\{h'(A, C), h'(B, C)\}.$$

Proof. By definition of h'.

Proposition 2.60. *Let* A_i, B_i *be subsets of* X *for all* $i \in I$, *then*

$$h(\cup_{i \in I} A_i, \cup_{i \in I} B_i) \leq \sup_{i \in I} h(A_i, B_i).$$

Proof. We have $h(\cup_i A_i, \cup_i B_i) = \max\{h'(\cup_i A_i, \cup_i B_i), h'(\cup_i B_i, \cup_i A_i)\}$. Take the first part: by Prop(s). 2.59 and 2.58 successively,

$$
\begin{aligned}
& h'(\cup_i A_i, \cup_i B_i) \\
= \quad & \sup_i h'(A_i, \cup_i B_i) \\
\leq \quad & \sup_i h'(A_i, B_i).
\end{aligned}
$$

Symmetrically for the second part, $h'(\cup_i B_i, \cup_i A_i) \leq \sup_i h'(B_i, A_i)$.

Now, we define contracting, expanding, and neutral relations regarded as multi-valued functions.

Definition 2.61 (Contracting, expanding, neutral relations). *A relation* f *defined on a domain* D *of a metric space* (X, d) *is contracting iff*

$$\exists r < 1, \forall x, y \in D, h(f(x), f(y)) \leq r \cdot d(x, y);$$

it is expanding iff

$$\exists r < 1, \forall x, y \in D, d(x, y) \leq r \cdot h(f(x), f(y));$$

it is neutral iff
$$\forall x, y \in D, d(x, y) = h(f(x), f(y)).$$

These relations are *variant* ones. Neutral relations are limits of expanding or contracting relations. Variant relations are opposed to constant relations previously defined.

Definition 2.62 (Variant relation). *A relation f is variant iff it is expanding, contracting, or neutral.*

The contractivity factor r of a variant relation (see Def. 2.61) generalizes the notion of Lipshitz coefficient in case of contractions.

Definition 2.63 (Contractivity factor). *For each variant relation f, the contractivity factor $\gamma(f)$ depends on the smallest positive real number $r < 1$ verifying the expressions of Def. 2.61:*

$$\gamma(f) = \begin{cases} r & \textit{iff} & f \textit{ is contracting,} \\ \frac{1}{r} & \textit{iff} & f \textit{ is expanding,} \\ 1 & \textit{iff} & f \textit{ is neutral.} \end{cases}$$

The contractivity factor remains valid when set-transformers are considered instead of multi-valued functions. This is due to the following proposition.

Proposition 2.64. *Let (X, f) a contracting RDS, then the corresponding set-transformer has the same contractivity factor.*

Proof. We prove the contracting case; the two other cases are left to the reader. By hypothesis, we know that there exists $r < 1$ such that $\forall x, y \in X$,

$$h(f(x), f(y)) \leq r \cdot d(x, y).$$

We want to prove that $\forall A, B \subseteq X$, we have also

$$h(f(A), f(B)) \leq r \cdot h(A, B).$$

By definition, $h(f(A), f(B)) = \max\{h'(f(A), f(B)), h'(f(B), f(A))\}$. Let us consider the first part, $h'(f(A), f(B))$: by Prop. 2.58, since $f(b) \subseteq f(B)$ by monotonicity (Prop. 2.35), we have $\forall b \in B$

$$h'(f(a), f(B)) \leq h'(f(a), f(b)).$$

From this, we have successively

$$h'(f(a), f(B)) \leq \inf_{b \in B} h'(f(a), f(b))$$

$$\sup_{a \in A} h'(f(a), f(B)) \leq \sup_{a \in A} \inf_{b \in B} h'(f(a), f(b))$$

$$h'(f(A), f(B)) \leq \sup_{a \in A} \inf_{b \in B} h(f(a), f(b))$$

$$\leq \sup_{a \in A} \inf_{b \in B} r \cdot d(a, b)$$

$$\leq r \cdot h'(A, B)$$

$$\leq r \cdot h(A, B).$$

The second part can be treated symmetrically.

Based on these three properties, the "kind" of a relation can be coded this way.

Definition 2.65 (Kind). *The kind, if any, of a relation f is defined by*

$$\kappa(f) = \begin{cases} - & \text{if } f \text{ is contracting} \\ 0 & \text{if } f \text{ is neutral} \\ + & \text{if } f \text{ is expanding.} \end{cases}$$

Remark 2.66. Many relations have no kind using this definition. Other cases can be seen from different viewpoints. For instance, suppose a relation f is defined on $X = Y \times Z$ instead of a single space Y, and it can be expressed as a product of independent relations defined on Y and Z respectively. Then, the kind can be defined globally on X or as a vector of kinds respectively defined on Y and Z. This second aspect will often be preferred, each time such a decomposition will be possible. The purpose of Chap. 3 is to introduce a family of composition operators on relations, allowing these considerations.

Finally, let us recall an important result from elementary calculus. A contracting function defined on a complete metric space has the interesting property to converge to a unique fixpoint in ω iterations, starting from any state of its domain. The following theorem is classical and sometimes called *Banach's fixpoint theorem*.

Theorem 2.67 (Contraction mapping theorem). *Let (X, d) be a complete metric space, and f be a contracting function defined on X, with contractivity factor γ. Then f has a unique fixpoint p, and $\forall x \in X$, the sequence $(f^k(x))_k$ converges to p. Moreover, $\forall x \in X, n, d(f^n(x), p) \leq \frac{c}{1-c} d(x, f(x))$.*

2.4 Transfinite Iterations

So far, we have restricted ourselves to unbounded finite iterations. Theorems have been presented to compute fixpoints of relations by successive approximations, provided some continuity assumptions (see §2.3.4). These assumptions can be weakened if transfinite iterations are allowed, i.e. containing more than any finite number of steps, and even more than ω steps. Using a transfinite iteration scheme based on ordinal numbers, monotonicity is sufficient to get convergence in the computation of fixpoints of functions defined on complete lattices. Despite the fact that some initial states do not entail monotonicity, a notion of limit can be defined ine the particular case of RDS. In the same way, transfinite trajectories are defined, too.

2.4.1 Motivation

Let us start with a simple example showing the usefulness of transfinite iterations: the function defined below is not a RDS since it is not closed, and convergence is obtained after more than ω iteration steps.

Example 2.68. Take a function defined as follows (see Fig. 2.6):

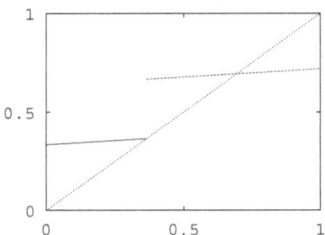

Fig. 2.6. Graph of $f(x)$ (and $y = x$, dotted line)

$$f(x) = \begin{cases} \frac{1}{12}x + \frac{1}{3} & \text{on } [0, \frac{4}{11}) \\ \frac{1}{12}x + \frac{7}{11} & \text{on } [\frac{4}{11}, 1]. \end{cases}$$

The part of f defined on $[0, \frac{4}{11})$ seems to have a fixed point in $\frac{4}{11}$ but there it is not defined. Actually, at this point, the right branch is defined, for which there is a true fixed point in $\frac{84}{121}$. Both branches of f are contracting, i.e. the contractivity factor is strictly smaller than 1 (here, it is equal to $\frac{1}{12}$).

   Thus, starting from any point in $[0, \frac{4}{11})$, ω iterations are needed to reach the "virtual" attracting fixed point $\frac{4}{11}$.

   For any initial point x_0, the nth iteration is given by $x_n = \frac{1}{12^n}(x_0 - \frac{4}{11}) + \frac{4}{11}$. This means that for x_n to be greater than a point $\frac{4}{11} - \frac{1}{m}$ very close to the virtual fixed point, n has to be in $\mathcal{O}(\log m)$. Precisely,

$$n \geq \left\lceil \frac{1}{\log 12}(\log m + \log(4 - 11x_0) - \log 11) \right\rceil.$$

If m tends to infinity, then n tends to infinity, too.

At the "virtual" fixpoint $\frac{4}{11}$, viz. after ω steps, a jump happens: the first branch should give $\frac{4}{11}$ but it is not defined, whereas the second branch gives $\frac{2}{3}$ and the iteration can keep going on this second branch. The fixed point is reached after ω other steps (this can be proved using the same argument as above). Hence, $\omega + \omega$ iterations are necessary to reach the fixed point of f.

From this example, we see that it can be interesting to have more than ω possible steps to reach a fixpoint.

Let us give another similar motivation. Using a finite deterministic automaton, it is clear that, from any state, a cycle is reached after a finite number of transitions called transient. Moreover, this transient contains a number of states which is at most equal to the number of states of the automaton. In order to let the system converge in any case, it is useful to define an iteration scheme permitting more steps than the number of states of the automaton. Repeating the argument with larger and larger automata that contain more and more states, we tend to infinity. This motivates the use of strictly infinite iteration schemes. Since we work with set-transformers, power sets are needed and we require possibly higher degrees of infinity. This motivates the use of transfinite iteration schemes.

2.4.2 Transfinite Fixpoint Theorem

The class of *ordinal numbers* (denoted $\mathbb{O}$) is well ordered by the classical "less than" relation $<$. The expression $\sqcup_{i \in I} n_i$ is used to denote the upper bound of the ordinals family $\{n_i \mid i \in I\}$. A *limit ordinal* n is such that $\sqcup_{k<n} k = n$. A *successor ordinal* n is such that $\sqcup_{k<n} k = n - 1$, where the predecessor of n is denoted by $n - 1$ [96]. In the following, we denote limit ordinals by $\mathbb{O}_l$ and successor ordinals by $\mathbb{O}_s$.

Let us assume that $\mathbb{L}(\leq, \bot, \top, \sqcap, \sqcup)$ is a complete lattice, with ordering relation $\leq$, bottom $\bot$, top $\top$, least upper bound operator $\sqcup$, and greatest lower bound $\sqcap$. With the same notations, $\eta_\mathbb{L}$ denotes the smallest ordinal number strictly greater than the cardinality of $\mathbb{L}$.

Now, we extend the notion of iteration to transfinite numbers. Limit elements of sequences are easy to define when these sequences are decreasing or increasing.

Definition 2.69 (Decreasing/increasing transfinite sequence). *A transfinite sequence $(X_i)_{i \in \mathbb{O}}$ is decreasing iff $\forall i \in \mathbb{O}_s, X_i \leq X_{i-1}$ and $\forall i \in \mathbb{O}_l, X_i = \sqcap_{j<i} X_j$. We denote it by $\downarrow_i X_i$.*

The same sequence is increasing iff $\forall i \in \mathbb{O}_s, X_{i-1} \leq X_i$ and $\forall i \in \mathbb{O}_l, X_i = \sqcup_{j<i} X_j$. We denote it by $\uparrow_i X_i$.

The concept of stability is straightforward.

Definition 2.70 (Stationary sequence). *The (decreasing or increasing) sequence $(X_i)_{i<\eta_\mathbb{L}}$ of elements of the complete lattice $\mathbb{L}$ is stationary iff $\exists n < \eta_\mathbb{L}, (m \geq n) \Rightarrow (X_n = X_m)$. The limit of this sequence is X_n.*

Based on this we can propose an evolution scheme: the basic case is $f^0(A) = A$; for all successor ordinals $n \neq 0 \in \mathbb{O}_s$, $f^n(A) = f(f^{n-1}(A))$; and for all limit ordinals $n \in \mathbb{O}_l$,

$$
\begin{aligned}
f^n(A) &= \sqcap_{m<n} f^m(A) \\
&\quad \text{if } \downarrow_m f^m(A) \\
&= \sqcup_{m<n} f^m(A) \\
&\quad \text{if } \uparrow_m f^m(A);
\end{aligned}
$$

finally, for all ordinals $n \neq 0 \in \mathbb{O}$, $f^{-n}(A) = (f^{-1})^n(A)$.

Continuity (see §§2.3.3, 2.3.4) entails stationarity in at most ω steps. The following proposition states the underlying equivalence [70].

Proposition 2.71. *Let $\mathbb{L}$ be a complete lattice. If $\eta_\mathbb{L} > \omega$, the sequence of successive iterates $(X_i)_{i<\eta_\mathbb{L}}$ is stable in at most ω steps iff f is and-continuous.*

Proof.

$$
\begin{aligned}
&\forall n < \eta_\mathbb{L}, (n \geq \omega) \Rightarrow (X_\omega = X_n) \\
\equiv\quad & X_\omega = X_{\omega+1} \\
\equiv\quad & X_\omega = f(X_\omega) \\
\equiv\quad & \sqcap_{i<\omega} X_i = f(\sqcap_{i<\omega} X_i) \\
\equiv\quad & \sqcap_{i<\omega} X_{i+1} = f(\sqcap_{i<\omega} X_i) \\
\equiv\quad & \sqcap_{i<\omega} f(X_i) = f(\sqcap_{i<\omega} X_i).
\end{aligned}
$$

Relaxing the assumption of converging in at most ω steps amounts to keeping monotonicity instead of continuity. The following theorem generalizes Knaster-Tarski's to a constructive fixpoint theorem for monotonic relations on complete lattices [70, 71].

Theorem 2.72 (Transfinite lattice fixpoint theorem). *Let $\mathbb{L}$ be a complete lattice. A decreasing iteration $(X_i)_{i<\eta_\mathbb{L}}$ starting from $A \geq f(A)$, and defined by a monotonic relation f, is a stationary decreasing sequence and its limit is $\mathbb{G}f$, the greatest fixed-point of f, $\leq A$.*

Similarly, an increasing iteration $(X_i)_{i<\eta_\mathbb{L}}$ starting from $A \leq f(A)$, and defined by a monotonic relation f, is a stationary increasing sequence and its limit is $\mathbb{S}f$, the least fixed-point of f, $\geq A$.

Remark 2.73. Most of the time, decreasing iterations are used, starting from the entire space X, which is the top element of the lattice. Increasing iterations starting from the bottom element do not help if the bottom is equal to the empty set, because of Prop(s). 2.21. Otherwise, an "undefined" relation based on an "undefined" element could do (see also Rem. 2.4), as well as working in another space, e.g. $\mathbb{K}(X)$ the nonempty compact subsets of X.

2.4.3 Transfinite Limits of Iterations

The iteration scheme on which Theorem 2.72 is based, is too restrictive because it happens that successive iterations from a set A are neither increasing nor decreasing. In this case, one would like to have notion of limit that remains compatible with these particular cases.

In terms of RDS, it is easy to define such a notion since every sequence in a compact set has accumulations points in this set. These accumulation points can serve as limit elements.

Definition 2.74 (Limit set). *Let A be a subset of X. The limit set of A by successive iterations of a RDS (X, f) is:*

$$f^\omega(A) = \cap_{i<\omega}\overline{\cup_{i\leq j<\omega} f^j(A)}.$$

Remark 2.75. − This notion is well defined: the result is always empty or closed, thanks to monotonicity and closure.
− Moreover, it reduces to $\cap_i \overline{f^i(A)}$ when $f(A) \subseteq A$ and to $\overline{\cup_i f^i(A)}$ when $A \subseteq f(A)$. Notice that these expressions are very similar to the limits of decreasing and increasing sequences (see Def. 2.69).

We generalize the notion to any complete lattice, which leads to the last iteration scheme.

Definition 2.76 (Transfinite iteration scheme). *Let $\mathbb{L}$ be a complete lattice, f a monotonic function on $\mathbb{L}$, and $A \in \mathbb{L}$. The transfinite iteration scheme is defined by: the basic case,*

$$f^0(A) = A;$$

for all successor ordinals $n \neq 0 \in \mathbb{O}_s$,

$$f^n(A) = f(f^{n-1}(A)),$$

and for all limit ordinals $n \in \mathbb{O}_l$,

$$f^n(A) = \sqcap_{i<n} \sqcup_{i\leq j<n} f^j(A).$$

finally, for all ordinals $n \neq 0 \in \mathbb{O}$,

$$f^{-n}(A) = (f^{-1})^n(A).$$

Remark 2.77. The lattice on which the evolution happens is supposed to be complete. Thus, any subset of the lattice has a greatest lower bound and a least upper bound. Consequently, the expression $\sqcap_{i<n} \sqcup_{i\leq j<n} f^j(A)$ is well defined. However, it is not necessarily compact when used in a compact space, because the infinite union is not always closed, even if A is closed.

Let us come back on Ex. 2.68, to illustrate how transfinite iterations can help.

Example 2.78 (Ex. 2.68 revisited). Let us pay attention to the definition of f, which does not belong to our general class of RDS, for f is not closed: $Dom(f) = [0,1]$ is closed but $Rg(f) = [\frac{1}{3}, \frac{4}{11}) \cup [\frac{2}{3}, \frac{95}{132}]$ is not closed. We could also define f as a relation by closing it: $f(\frac{4}{11}) = \{\frac{4}{11}, \frac{2}{3}\}$. In case of a particular evolution starting from a state of the first interval $[0, \frac{4}{11})$, $\omega + \omega$ iterations are necessary to reach $\{\frac{4}{11}, \frac{84}{121}\}$.

Using transfinite iterations, the strong assumptions of Def. 2.6 can be weakened to all relations in almost all spaces. Set-transformers are always monotonic, which is sufficient to guarantee the convergence of successive iterations. The trade-off consists in waiting possibly more than any finite number of time, and even more than ω steps.

Finally, we have to discuss how to extend the notion of nondeterministic dynamics (Def. 2.9) to transfinite sequences. We use accumulation points again.

Definition 2.79 (Transfinite nondeterministic forward dynamics).
The transfinite nondeterministic forward dynamics of a RDS (X, f) from a set $A \subseteq X$ of initial conditions is

$$\theta(A, f) \;=\; \{s \in X^{\mathbb{O}} \mid (s_0 \in A)$$
$$\wedge (\forall n \neq 0 \in \mathbb{O}_s, (s_{n-1}, s_n) \in f)$$
$$\wedge (\forall n \in \mathbb{O}_l, s_n \in \sqcap_{i<n} \sqcup_{i\leq j<n} \{s_j\})\}.$$

2.5 Discussion

In this section, we compare important aspects of our work with related notions: the relational framework we develop, as compared to the classical functional view; set-level dynamics and predicate-transformers; point-level dynamics and trace semantics; explicit nondeterminism and probabilistic choices; transfinite iterations; generalized time structures.

2.5.1 Relations vs Functions

Using relations is frequent in the theory of programs but not in the field of dynamical systems, though some authors have proposed to introduce relations as fundamental dynamical systems.

In particular, in [9], the author builds different relations on top of functions, in order to express several variants of recurrence, and suggests the introduction of relations at the basic level in order to get a homogeneous treatment of systems.

In [221], relations are not introduced explicitly; the authors allow the superposition of several functions, leading to nondeterminism, through difference and differential inequations and inclusions.

In a more abstract way, noninvertible dynamical systems are related to semi-groups, while invertibility confers them the full power of groups. In both cases, many interesting results come directly from group theory, when such relationships are established. For example, in [37], the author makes use of group theory in Rubik's cube (!) to illustrate chaos in finite (but large) spaces.

2.5.2 Set-Level Dynamics and Predicate-Transformers

Set-transformers are not new. They are used in general topology [9] and fractal theory [328, 159, 140, 28, 325].

When sets are specified by predicates, set-transformers are expressed as *predicate-transformers*. In program theory, these predicate functions are used to express the semantics of programs (e.g. [91, 150, 245, 93, 246]) and transition systems in general (e.g. [284]).

The interesting relationships between relations, predicate-transformers, multi-valued functions, and their algebraic construction have been investigated in [39, 112].

We summarize below the equivalences between set-transformers and existing operators:

$$\begin{array}{ccccc} R & \equiv & R_+ & \equiv & post[R] \\ R^{-1} & \equiv & R_- & \equiv & pre[R] \\ \text{[this monograph]} & & \text{[286]} & & \text{[284]} \end{array}$$

and

$$\begin{array}{cll} & wp \cdot R \cdot A & \text{[91]} \\ \equiv & \mathcal{WR} \cdot R \cdot A & \text{[246]} \\ \equiv & pre[R](A) \wedge \neg pre[R](\neg A) & \text{[284]} \\ \equiv & R^{-1}(A) \cap X \backslash R^{-1}(X \backslash A) & \text{[this monograph]} \end{array}$$

assuming that a unique sink can be reached by R when non-termination is possible, according to [150, 130].

Some authors directly consider programs as relations. This allows them to express program properties and specifications in a very structured and clear algebraic way; see for instance [130, 150, 152, 246, 313]. In this case, the basic objects on which structured systems are built are nothing but relations, as in our approach.

After Dijkstra's work, predicate-transformers have been extended to parallel programs [61, 194, 62], and to probabilistic programs [234, 237, 236].

In control theory , some dynamical systems have been recently analyzed using predicate-transformers [187]. However, they do not use a compositional analysis as the one we develop in the next chapters.

2.5.3 Point-Level Dynamics and Trace Semantics

We have introduced two equivalent notions to describe the evolutions of systems: a set-level dynamics based on successive iterations of set-transformers, and a point-level dynamics based on sets of (possibly nondeterministic) trajectories.

Defining systems through their trajectories is not new. In dynamical systems theory, the notion of generating system can be abandoned without loss of generality [21, 327, 221, 288]. In program theory, traces define the semantics of sequential or parallel programs [15, 216, 189, 309, 67, 89].

2.5.4 Nondeterminism and Probabilistic Choices

Relations intrinsically contain nondeterminism. A function associates at most one state to every state of its domain. A relation can associate a whole set of states to every state of its domain. The dynamics chosen to describe the evolution thus plays an important role in this context: it can be deterministic, and take all images as the result of the evolution; it can be probabilistic, and select one of the images using a probability density; or, it can be nondeterministic, and arbitrarily pick one of the possible images. When nondeterminism is preferred, no quantitative assumption can be made on the performed choices.

If A is a subset of X, then $!(A)$ represents a *nondeterministic choice* of any element $a \in A$. Thus, for some $a \in A$, $!(A) = a$. Such an operation is discussed in [32], and related to a nondeterministic choice operator introduced by Hilbert [149].

This choice operator could be composed with set-transformers. Considering a relation f of $\mathcal{R}(X)$ and a set $A \subseteq X$, we can choose one element of the result:

$$f^!(A) \quad = \quad !(\cup_{u \in A} f(u)).$$

From a pragmatic point of view, this operator ! is sometimes very useful because it allows to consider one state per iteration instead of a set of states. For instance, using Iterated Function Systems [28] under fairness-like conditions, the following equality is verified asymptotically:

$$\overline{\cup_n f^{!n}(A)} = \overline{\cup_n f^n(A)}.$$

These choice-based iterations are also called "pseudo-periodic chaotic iterations" in [274].

A probabilistic choice between several images of a state or a set of states can also be introduced. Some authors have proposed theoretical possibilities extending the framework of iterated dynamical systems to nondeterminacy, without loosing quantitative properties [28, 97, 160]. This leads to higher levels of description (i.e. abstractions) than the set level: measures and random measures.

Finally, nondeterminism and probabilities can be related, like in the semantics of programs. Among others, the work of [162, 234, 237, 236] establishes interesting results that could be introduced in our framework.

2.5.5 Transfinite Iterations

We have presented several discrete-time evolution schemes of relations by means of successive iterations. Finite, infinite and transfinite iterations are possible; their convergence generally relies on fixpoint theorems.

To our knowledge, the use of transfinite iterations as general iteration scheme in the context of dynamical systems is quite rare. The authors usually consider continuous functions defined on compact metric spaces, where we know that at most ω iterations are needed to compute fixpoints (see §2.4).

However, lattice fixpoint theorems provide convergence results using only monotonic relations. Their constructive version requires continuity or monotonicity and transfinite iterations.

Transfinite iterations have been introduced in the context of program semantics to guarantee convergence of iterative computation of solutions of fixpoint equations without adding the stronger assumption of continuity of the operators involved [70, 71].

At this point, we conjecture that contractions on complete metric spaces and continuous functions on complete partial orders are instances of a same abstract pair function–space. Category theory should provide us the tools to investigate this open question. First steps toward an unification of these models have been studied in [324, 40, 186, 5, 297]. We hope to study this interesting question later, to relate subsequent fixpoint theorems based on these models, and express them as instances of an abstract fixpoint theorem. As subgoal, we would like to propose a transfinite version of the contraction mapping theorem, and analyze the conditions under which the result could remain valid by weakening some assumptions and keeping the transfinite iteration scheme. In short, the question is:

> *What is the equivalent of weakening relations from continuity to monotonicity in the context of complete metric spaces?*

2.5.6 Time Structure

Up to now, we have considered discrete-time dynamics based on successive iterations. We could abstract the evolution scheme using an *evolution operator*, and map a relation into a set-transformer describing a specific kind of dynamics, continuous or discrete in time. We could also generalize time to any other ordered structure.

In particular, the point-level nondeterministic dynamics presented in this chapter can also be generalized to other time structures. Each component of the dynamics, i.e. a trajectory, is a sequence of states, and can be seen as a function defined from time to space. We have presented the discrete-time version of this dynamics, viz. time was always $\mathbb{N}$, $\mathbb{Z}$, or $\mathbb{O}$, but other time domains could be used. For instance, each trajectory of the dynamics of a system can be defined on $\mathbb{R}$, and describe continuous-time evolutions of systems [221, 326].

Finally, mixing different types of evolutions in a same framework would be interesting in order to model real-time systems where discrete-time machines interact with continuous-time events from the environment. This is the goal of hybrid systems theory, where systems based on different time structures are composed together [320, 16, 287, 288].

3. Dynamics of Composed Relations

In Chap. 2, we defined relational dynamical systems and different ways to express their discrete-time evolution. In this chapter, we introduce composed dynamical systems. By means of composition operators, structured relations are constructed from basic ones. The first step toward our objective is the analysis of the set-level and point-level dynamics of systems by composition: global (resp. individual) properties are simply the dynamics of composed systems (resp. components).

The chapter is organized as follows: after a short introduction in §3.1, we present composition operators on relational dynamical systems in §3.2; then, in §3.3, we study some important dynamical aspects of composed relations: we analyze how composition propagates from the structure of relational systems to set-level and point-level dynamics; in §3.4, we detail algebraic properties of composition operators, and give composition laws of compositions operators; these multiple compositions naturally lead to fixpoint considerations; finally, in §3.5, we close the chapter with a discussion.

3.1 Structural Composition

Decomposing a mathematical object is often crucial to understand it. Let us consider a basic example, using whole numbers, for which several representations can be chosen, depending on the type of information one wants to get:

$$17 \quad = \quad \underbrace{succ(succ(\cdots succ(0)\cdots))}_{17 \text{ times}}$$

$$= \quad 7^{\text{th}} \text{ prime number}$$

$$= \quad 1 \times 10 + 7.$$

In program theory, sequential and parallel programs are considered as structurally composed systems. From simple composition operators like sequential composition of guarded-commands in Dijkstra's language [91] to modular approaches in software engineering, composition is of fundamental importance. Parallel composition, "rendez-vous" synchronization, product,

F. Geurts: Abstract Compositional Analysis of Iterated Relations, LNCS 1426, pp. 53-79, 1998.
© Springer-Verlag Berlin Heidelberg 1998

sum, are good examples of structuring means of programs, transition systems, or algebraic processes [151, 216, 222, 3]. Compositional analyses of these systems are elaborated, to overcome the impossibility of dealing with huge monolithic systems.

Surprisingly, dynamical systems are often studied as complex mathematical objects, without trying to decompose them into simpler components. Some exceptions exist, particularly in the field of hybrid systems [244]. Usually, composition designates successive applications of functions to given states: f and g being two functions defined on X, $x \in X$, $(f \circ g)(x) = f(g(x))$. Another structural way to combine systems is the simple Cartesian product of function; for instance, $(f \times g)(x, y) = (f(x), g(y))$. Finally, independent variables can be mixed together, as in any matrix-vector multiplication.

We now present composition operators on relational dynamical systems, inspired by classical operators from relation algebra, dynamical systems and program theories: inversion (already encountered in Chap. 2), domain and range restrictions, negation, difference, intersection, union (nondeterministic choice whenever possible), free product (without interaction between components, as in the Cartesian product) and connected product (with explicit interaction).

Our objective is to extend "interesting" properties from simple systems, viz. easy to analyze, to composed systems obtained by combination of these individual ones. Interesting properties can vary from one-step evolution to complete dynamics, invariance, attraction, to computational characteristics, etc. However, this chapter focuses on set- and point-level dynamics. Other properties will be examined later on (see Chap(s). 6–9).

3.2 Composition of Relations

We have presented relational dynamical systems in the previous chapter (see Def. 2.6). Now, we introduce composition operators on relations, and repeat the (iterative) dynamical construction in addition to the structural combination. We turn our attention to the recursive construction of new relations from basic ones, using composition operators. These new relations can in turn serve to produce other relations.

Almost all operators that we propose here can be found in the literature, for many of them are classical set-theoretic operations and relational operations. However, for the sake of completeness, we precisely define them and we illustrate their use by means of simple examples.

We start with basic cases of composition, involving single relations: unary operators. After that, we define some operators taking at least two relations as arguments.

3.2.1 Unary Operators

The first operators we use only involve one relation as argument. Among them, we find inversion, domain restriction, range restriction, and negation.

Definition 3.1 (Inversion). *The inverse of a relation $f \in \mathcal{R}(X, Y)$ is given by*

$$\begin{array}{rl} ^{-1} \quad : & \mathcal{R}(X,Y) \mapsto \mathcal{R}(Y,X) \\ s.t. & f^{-1} = \{(v,u) \mid (u,v) \in f\}. \end{array}$$

Example 3.2. Let us define the following relation $f_1 = \{(1,a),(2,b),(3,a),(4,c)\}$ in $\mathcal{R}(\{1,2,3,4\},\{a,b,c\})$. Its inverse is

$$f_1^{-1} = \{(a,1),(a,3),(b,2),(c,4)\}.$$

Domain restriction is an operator restricting the set of states to which the relation can be applied.

Definition 3.3 (Domain restriction). *The domain restriction of a relation $f \in \mathcal{R}(X,Y)$ to $B \subseteq X$ is*

$$\begin{array}{rl} (B \rightarrow \ ) \quad : & \mathcal{R}(X,Y) \mapsto \mathcal{R}(X,Y) \\ s.t. & (B \rightarrow f) = f \cap (B \times Y). \end{array}$$

Example 3.4. We consider relation f_1 of Ex. 3.2, and restrict its domain to $\{1,2\}$. This gives

$$(\{1,2\} \rightarrow f_1) = \{(1,a),(2,b)\}.$$

Range restriction is the dual operator of domain restriction: it restricts the set of images a relation can range in.

Definition 3.5 (Range restriction). *The range restriction of a relation $f \in \mathcal{R}(X,Y)$ to $B \subseteq Y$ is*

$$\begin{array}{rl} (\ \leftarrow B) \quad : & \mathcal{R}(X,Y) \mapsto \mathcal{R}(X,Y) \\ s.t. & (f \leftarrow B) = f \cap (X \times B). \end{array}$$

Example 3.6. Restricting the range of relation f_1 of Ex. 3.2 to $\{a,b\}$ gives

$$(f_1 \leftarrow \{a,b\}) = \{(1,a),(2,b),(3,a)\}.$$

Notation 3.7. *The conjunction of domain and range restrictions is denoted by*

$$(A \rightarrow f \leftarrow B) = (A \rightarrow (f \leftarrow B)) = ((A \rightarrow f) \leftarrow B).$$

We introduce a last unary operator, negation, based on set difference.

Definition 3.8 (Negation). *The negation or complement of a relation $f \in$* $\mathcal{R}(X,Y)$ *is given by*

$$\sim \quad : \quad \mathcal{R}(X,Y) \mapsto \mathcal{R}(X,Y)$$
$$\text{s.t.} \quad \sim f = (X \times Y) \backslash f.$$

Example 3.9. Let us compute the negation of relation f_1 defined in Ex. 3.2. This gives:

$$\sim f_1 \;\; = \;\; \{1,2,3,4\} \times \{a,b,c\} \backslash f_1$$
$$= \;\; \{(1,b),(1,c),(2,a),(2,c),(3,b),(3,c),(4,a),(4,b)\}.$$

3.2.2 N-Ary Operators

Now, we come to (at least) binary operators. Sequential composition allows to compose several relations as a classical functional composition (e.g. $f(g(x))$). Intersection, union and difference immediately extend set-theoretic operations. This is simply due to the fact that relations are nothing but sets. Finally, two kinds of product are introduced: a Cartesian product of relations (and their respective spaces) called "free" because no interaction exists between components; a "connected" version of this product, where explicit interaction is added between components.

The sequential composition of two binary relations is given below.

Definition 3.10 (Sequential composition). *The sequential composition of two relations $f \in \mathcal{R}(X,Y)$ and $g \in \mathcal{R}(Y,Z)$ is given by*

$$; \quad : \quad \mathcal{R}(X,Y) \times \mathcal{R}(Y,Z) \mapsto \mathcal{R}(X,Z)$$
$$\text{s.t.} \quad f;g = \Pi_{1,3}((f \times Z) \cap (X \times g)).$$

Example 3.11. Let us introduce another relation on $\mathcal{R}(\{a,b,c,d\},\{\alpha,\beta,\gamma,\delta,\epsilon\})$, $f_2 = \{(a,\alpha),(b,\gamma),(c,\delta),(d,\epsilon),(a,\beta)\}$. Composing f_1 of Ex. 3.2 with f_2 sequentially gives

$$f_1;f_2 = \{(1,\alpha),(1,\beta),(2,\gamma),(3,\alpha),(3,\beta),(4,\delta)\}.$$

The intersection of relations expresses the classical set intersection: we consider *common domains* and take a result only if it belongs to the common range of the relations involved.

Definition 3.12 (Intersection). *The intersection of two relations f and g of $\mathcal{R}(X,Y)$ is their set-intersection:*

$$\cap \quad : \quad \mathcal{R}(X,Y) \times \mathcal{R}(X,Y) \mapsto \mathcal{R}(X,Y)$$
$$\text{s.t.} \quad f \cap g = \{(x,y) \mid ((x,y) \in f) \wedge ((x,y) \in g).$$

Example 3.13. Here,   we   introduce   a   new   relation   on $\mathcal{R}(\{1,2,3,4,5\},\{a,b,c,d\})$: f_3 $=$ $\{(1,a),(2,b),(3,d),(4,c),(5,a)\}$. Its intersection with relation f_1 of Ex. 3.2 is

$$f_1 \cap f_3 = \{(1,a),(2,b),(4,c)\}.$$

Example 3.14. In Fig. 3.1, two very simple functions on $[0,1]$ are represented. Their overlapping subspace is $[\frac{1}{4},\frac{3}{4}] \times [\frac{1}{4},\frac{7}{8}]$. Their interesection is defined on this subspace as the pairs (state, image) they share.

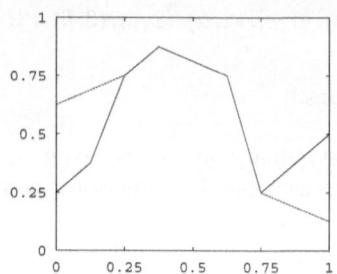

Fig. 3.1. Intersection of two functions

The union of relations expresses the classical notion of set union. It emphasizes a possible nondeterministic choice between different systems.

Definition 3.15 (Union). *The union of two relations* f *and* g *of* $\mathcal{R}(X,Y)$ *is their set-union:*

$$\cup \ : \ \mathcal{R}(X,Y) \times \mathcal{R}(X,Y) \mapsto \mathcal{R}(X,Y)$$
$$s.t. \quad f \cup g = \{(x,y) \mid ((x,y) \in f) \vee ((x,y) \in g)\}.$$

Example 3.16. Let us consider f_1 of Ex. 3.2 and f_3 of Ex. 3.13. Their union is given by

$$f_1 \cup f_3 = \{(1,a),(2,b),(3,a),(3,d),(4,c),(5,a)\}.$$

The difference of relations is again straightforward. This operator is not commutative. It is equivalent to the classical notion of set-difference.

Definition 3.17 (Difference). *The difference of two relations* f *and* g *of* $\mathcal{R}(X,Y)$ *is their set-difference:*

$$\backslash \ : \ \mathcal{R}(X,Y) \times \mathcal{R}(X,Y) \mapsto \mathcal{R}(X,Y)$$
$$s.t. \quad f \backslash g = \{(x,y) \mid ((x,y) \in f) \wedge ((x,y) \notin g)\}.$$

Example 3.18. We consider the same relations as in Ex. 3.16. Two differences can be computed:

$$f_1 \backslash f_3 = \{(3,a)\} \text{ and } f_3 \backslash f_1 = \{(3,d),(5,a)\}.$$

The idea of the next operator is the synchronous parallelization of several relations, without any other interaction between them.

Definition 3.19 (Free product). *The free product of two relations* $f \in \mathcal{R}(X)$ *and* $g \in \mathcal{R}(Y)$ *is given by*

$$
\begin{aligned}
\times \; : \quad & \mathcal{R}(X) \times \mathcal{R}(Y) \mapsto \mathcal{R}(X \times Y) \\
\text{s.t.} \quad & f \times g = \{((u,v),(x,y)) \mid (u,x) \in f \wedge (v,y) \in g\}.
\end{aligned}
$$

Remark 3.20. Notice our definition does not consider "flat" products: the result is a set of pairs of pairs in this particular case, i.e. $\{((u,v),(x,y))\}$, instead of a "flat" set of quadruples, i.e. $\{(u,x,v,y)\}$ or $\{(u,v,x,y)\}$. This means that the free product is neither associative nor commutative. It can then be used to create hierarchical systems. Of course, as mentioned earlier, the free product is not necessarily binary, its arity can be as large as we want. For example, to create a flat three-dimensional system, we just need the ternary product instead of a composition of two binary products:

$$
\begin{aligned}
& f_1 \times f_2 \times f_3 \\
= \; & \{(u,x)\} \times \{(v,y)\} \times \{(w,z)\} \\
= \; & \{((u,v,w),(x,y,z))\} \\
\neq \; & \{((u,(v,w)),(x,(y,z)))\} \\
= \; & \{(u,x)\} \times \{((v,w),(y,z))\} \\
= \; & \{(u,x)\} \times (\{(v,y)\} \times \{(w,z)\}) \\
= \; & f_1 \times (f_2 \times f_3) \\
\neq \; & (f_1 \times f_2) \times f_3.
\end{aligned}
$$

Sometimes, two operators are used to define these two compositions, namely "structure construction" and "product". To distinguish between these two possible interpretations, we will explicitly use parentheses with the structure constructor.

Example 3.21. We consider relation f_1 of Ex. 3.2 and relation $f_3 \backslash f_1$ obtained in Ex. 3.18. Their free product is

$$
\begin{aligned}
& f_1 \times (f_3 \backslash f_1) \\
= \; & \{((1,3),(a,d)),((2,3),(b,d)),((3,3),(a,d)),((4,3),(c,d)), \\
& ((1,5),(a,a)),((2,5),(b,a)),((3,5),(a,a)),((4,5),(c,a))\}.
\end{aligned}
$$

Now, we keep the parallel execution of several relations present in the free product but we add interactions between components. Many different kinds of interactions can be proposed and this operator is in fact the most useful in practice. We begin with a fairly general definition: each component relation has in itself the possibility of several influencing factors, i.e. it can have more than just one argument, and a general relation describing the neighborhood of each component has to be provided. This leads to the definition below, where three principal components appear: a global space, a family of local neighborhoods, and a local relation defined for each one of them. We isolate the arguments of each local relation using the projection Π_R, where R plays a role of dependency relation between variables or subspaces, and we compute these local relations individually using the free product $\times_{i \in J} g_i$.

Definition 3.22 (Connected product). *Let J be a set of indices, $E = \times_{i \in J} X_i$ be the global space, i.e. the Cartesian product of individual state spaces, $R \in \mathcal{R}(J)$ and $g_i \in \mathcal{R}(\times_{j \in R(i)} X_j, X_i)$ be relations, then the connected product is given by*

$$\otimes_R \quad : \quad \times_{i \in J} \mathcal{R}(\times_{j \in R(i)} X_j, X_i) \mapsto \mathcal{R}(\times_{i \in J} X_i)$$
$$s.t. \qquad \otimes_R g_{i \in J} = (\Pi_R); (\times_{i \in J} g_i).$$

Example 3.23. To illustrate the connected product in a simple manner, we consider Hénon's map [139]:

$$H(x, y) = (1 - \mu x^2 + y, bx).$$

This can be rewritten as a connected product based on a connection relation $R = \{(1, 1), (1, 2), (2, 1)\}$, where 1 (resp. 2) represents the first (resp. second) variable x (resp. y). The pair $(2, 2)$ is not present because the second term bx does not depend on y. Two local functions are defined: $g_1(x, y) = 1 - \mu x^2 + y$ and $g_2(z) = bz$. Hénon's map is then equal to $g_1 \otimes_R g_2$.

Let us finally illustrate this definition with another classical example: we show that cellular automata can be rewritten as connected products. In Chap. 8, we will come back on these automata, and study them using the tools developed here after.

Example 3.24 (Cellular automaton). A one-dimensional bi-infinite binary cellular automaton is defined as follows: $J = \mathbb{Z}$ is the lattice of cells, $R = \{(i, i - 1), (i, i), (i, i + 1) \mid i \in J\}$ describes the neighborhood of each cell, and for all $i \in J$, $X_i = \{0, 1\}$ is the local state space, and $g_i = g \in \mathcal{R}(X^3, X)$ is the local transition function. Then, the automaton is completely characterized by the connected product $\otimes_R g$.

3.2.3 Composed Dynamical Systems

Based on the composition operators introduced above, we can define a notion of composed or structured dynamical system.

Definition 3.25 (Composed dynamical system). *A composed dynamical system is an arbitrary composition of relational dynamical systems (Def. 2.6), using the operators: inversion (Def. 3.1), domain and range restrictions (Def(s). 3.3 and 3.5), negation (Def. 3.8), difference (Def. 3.17), intersection (Def. 3.12), union (Def. 3.15), free product (Def. 3.19), and connected product (Def. 3.22).*

Remark 3.26 (RDS-preserving compositions). The arbitrary composition of RDS is not always a RDS. Let us review each operator separately.

- Inversion: always correct, by Prop. 2.30.
- Domain and range restrictions: valid if the restricting set is closed, as any intersection of closed sets is closed, and restrictions are based on intersection.
- Negation: generally not valid because the complement of a closed set is open and, thus, it is not a RDS. On the other hand, it can be correct when the topology is the power set of the underlying space for instance, since all sets are both closed and open.
- Sequential composition: always valid as projection on compact spaces and intersection preserve closed sets.
- Intersection: always valid, since arbitrary intersections of closed sets are closed.
- Union: finite composition only, apart from special cases as the one we mentioned for negation.
- Difference: same problem as the one we mentioned for negation.
- Free product: always valid, since any Cartesian product of closed sets is closed.
- Connected product: always valid, since projection and free products preserve closed sets.

In general, one is interested by decomposing a RDS into simpler components, that is, relational systems which are easy to analyze. For instance, variant relations (see Def. 2.62) are simple enough to be considered as basic systems.

A few examples illustrate the decomposition of some classical relations.

Example 3.27. The logistic map $f(x) = 4x(1 - x)$ defined on $[0, 1]$ (see Fig. 3.2) is the union of two simple injective relations $f_1 = ([0, \frac{1}{2}] \to f)$ and $f_2 = ([\frac{1}{2}, 1] \to f)$: $f = f_1 \cup f_2$.

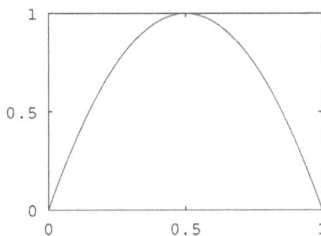

Fig. 3.2. Graph of $f(x) = 4x(1 - x)$

The function is contracting between $\frac{3}{8}$ and $\frac{5}{8}$, and expanding otherwise. It is thus also possible to rewrite it as a union of four injective simple variant relations.

The inverse of f, f^{-1}, has the same properties: it can be decomposed into two simple injective branches, respectively inverses of f_1 and f_2: $f^{-1} = f_1^{-1} \cup f_2^{-1}$.

Example 3.28. Another type of function is the family $g(x) = nx \bmod 1$ defined on $[0, 1]$ with $n \in \mathbb{N}$. Each one of them (see e.g. Fig. 3.3) is again a finite union of simple injective variant relations. Inverses of these functions are also

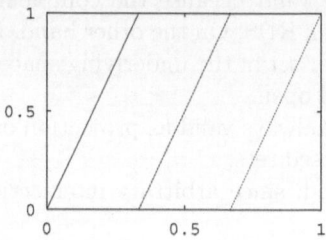

Fig. 3.3. Graph of $g(x) = 3x \bmod 1$

finite unions of simple injective variant relations.

Example 3.29. An interesting function is $h(x) = (\sin(\frac{1}{x}) \leftarrow [0, 1])$ (see Fig. 3.4) because of its infinitely many branches. In fact, there is only a countable

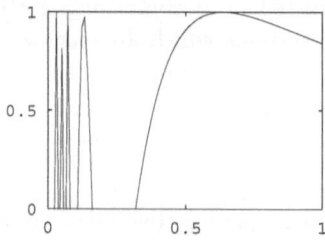

Fig. 3.4. Graph of $h(x) = (\sin(\frac{1}{x}) \leftarrow [0, 1])$

infinity of branches, and each one of them is a simple injective relation. Here, we thus have a countable union of basic relations. The inverse follows the same structure.

For it to be useful, the decomposition has to be composed of at most a (possibly finite) countable number of basic relations. Here we have shown examples of countable unions, but the same can be developed using other operators, like restrictions, sequential composition, intersection, or products.

3.3 Dynamics of Composed Relations

In Chap. 2, we have defined relational discrete-time dynamical systems and, in §3.2, we have presented how we can build composed systems from simpler ones.

The main objective of this monograph is the study of dynamical and computational properties of composed dynamical systems by the adequate combination of individual analyses of their components. To this end, since dynamical and computational properties both rely on the dynamics of systems, we have to study how structural composition propagates through evolution. This is the aim of this section: first, we determine one-step set-level evolution of composed systems; then, we study the impact of composition on the point-level dynamics.

3.3.1 One-Step Set-Level Evolution of Composed Relations

We now analyze properties of set-transformers for composed relations.

Proposition 3.30 (One-step iteration of composed relations). *Let* $f, g \in \mathcal{R}(X)$, $A \subseteq X$ *and* $a \in X$. *Then, the following statements hold:*

$$
\begin{align}
(B \to f)(A) &= f(A \cap B), \tag{3.1} \\
(f \leftarrow B)(A) &= f(A) \cap B, \tag{3.2} \\
\sim f(a) &= X \backslash f(a), \tag{3.3} \\
\sim f(A) &= X \backslash \cap_{a \in A} f(a), \tag{3.4}
\end{align}
$$

$$
\begin{align}
f \backslash g(A) &= f(A) \backslash f \cap g(A) \tag{3.5} \\
&\supseteq f(A) \backslash g(A), \\
f \cup g(A) &= f(A) \cup g(A), \tag{3.6} \\
f \cap g(A) &\subseteq f(A) \cap g(A), \tag{3.7} \\
f \times g(A) &\subseteq f(\Pi_1(A)) \times g(\Pi_2(A)), \tag{3.8} \\
f; g(A) &= g(f(A)), \tag{3.9} \\
\otimes_R g_i(A) &\subseteq \times_i g_i(\Pi_{R(i)}(A)). \tag{3.10}
\end{align}
$$

Proof. – Proof of (3.1):

$$
\begin{align}
(B \to f)(A) &= \{v \mid \exists u \in A, (u, v) \in (B \to f)\} \\
&= \{v \mid \exists u \in A, (u, v) \in f \cap B \times X\} \\
&= \{v \mid \exists u \in A \cap B, (u, v) \in f\} \\
&= f(A \cap B).
\end{align}
$$

– Proof of (3.2):

$$
\begin{aligned}
(f \leftarrow B)(A) \ &= \ \{v \mid \exists u \in A, (u,v) \in (f \leftarrow B)\} \\
&= \ \{v \mid \exists u \in A, (u,v) \in f \cap X \times B\} \\
&= \ \{v \mid \exists u \in A, (u,v) \in f\} \cap B \\
&= \ f(A) \cap B.
\end{aligned}
$$

– Proof of (3.3):

$$
\begin{aligned}
\sim f(a) \ &= \ (X \times X \backslash f)(a) \\
&= \ X \backslash f(a).
\end{aligned}
$$

– Proof of (3.4):

$$
\begin{aligned}
\sim f(A) \ &= \ (X \times X \backslash f)(A) \\
&= \ \cup_{a \in A}(X \times X \backslash f)(a) \\
&= \ \cup_{a \in A}(X \backslash f(a)) \\
&= \ X \backslash \cap_{a \in A} f(A).
\end{aligned}
$$

– Proof of (3.5):

$$
\begin{aligned}
f \backslash g(A) \ &= \ \{v \mid \exists u \in A, (u,v) \in f \wedge (u,v) \notin g\} \\
&= \ \{v \mid \exists u \in A, (u,v) \in f\} \backslash \{v \mid \exists u \in A, (u,v) \in f \cap g\} \\
&= \ f(A) \backslash f \cap g(A) \\
&\supseteq \ f(A) \backslash g(A).
\end{aligned}
$$

– Proof of (3.6):

$$
\begin{aligned}
f \cup g(A) \ &= \ \{v \mid \exists u \in A, (u,v) \in f \cup g\} \\
&= \ \{v \mid \exists u \in A, (u,v) \in f \vee (u,v) \in g\} \\
&= \ \{v \mid \exists u \in A, (u,v) \in f\} \cup \{v \mid \exists u \in A, (u,v) \in g\} \\
&= \ f(A) \cup g(A).
\end{aligned}
$$

– Proof of (3.7):

$$
\begin{aligned}
x \in f \cap g(A) \ &\equiv \ \exists u \in A, (u,x) \in f \wedge (u,x) \in g \\
&\Rightarrow \ \exists u \in A, (u,x) \in f \wedge \exists u \in A, (u,x) \in g \\
&\equiv \ x \in f(A) \wedge x \in g(A) \\
&\equiv \ x \in f(A) \cap g(A).
\end{aligned}
$$

To show that the reverse implication is not always verified, a simple example is sufficient. Consider two relations in the plane $\mathbb{R}^2$. Let f be the closed segment going from $(0,1)$ to $(1,3)$, and g be defined by the closed segment $(0,2) - (1,0)$. Their intersection is in $(\frac{1}{4}, \frac{3}{2})$. Let A be the whole interval $[0,1]$. We have $f \cap g(A) = \{\frac{3}{2}\}$, while $f(A) \cap g(A) = [1,2]$.

– Proof of (3.8):

$$\begin{aligned}
x \in f \times g(A) &\equiv \exists u = (u_1, u_2) \in A, (u, x) \in f \times g \\
&\equiv \exists (u_1, u_2) \in A, (u_1, x_1) \in f \wedge (u_2, x_2) \in g \\
&\Rightarrow \exists u_1 \in \Pi_1(A), (u_1, x_1) \in f \\
&\quad \wedge \exists u_2 \in \Pi_2(A), (u_2, x_2) \in g \\
&\equiv x_1 \in f(\Pi_1(A)) \wedge x_2 \in g(\Pi_2(A)) \\
&\equiv x \in f(\Pi_1(A)) \times g(\Pi_2(A)).
\end{aligned}$$

To show that the converse is not always verified, let us give a simple example. Take $A = \{(1, 2), (3, 4)\}$. The projections are $\Pi_1(A) = \{1, 3\}$ and $\Pi_2(A) = \{2, 4\}$. Let the relations be $f = \{(1, 4), (3, 7)\}$ and $g = \{(2, 1), (4, 3)\}$. Let us compute $f \times g(A) = \{(4, 1), (7, 3)\}$, $f(\Pi_1(A)) = \{4, 7\}$, and $g(\Pi_2(A)) = \{1, 3\}$. Then $(4, 3)$ does not belong to the first set while it belongs to the Cartesian product of the last two ones.

– Proof of (3.9):

$$\begin{aligned}
f; g(A) &= \{w \mid \exists u \in A, (u, w) \in f; g\} \\
&= \{w \mid \exists u, v, u \in A \wedge (u, v) \in f \wedge (v, w) \in g\} \\
&= \{w \mid \exists v, \exists u \in A, (u, v) \in f \wedge (v, w) \in g\} \\
&= \{w \mid \exists v \in f(A), (v, w) \in g\} \\
&= g(f(A)).
\end{aligned}$$

– Proof of (3.10):

$$\begin{aligned}
\otimes_R g_i(A) &= \cup_{u \in A} \otimes_R g_i(u) \\
&= \cup_{u \in A} (\Pi_R; \times_i g_i)(u) \\
&= \cup_{u \in A} \times_i g_i((\Pi_{R(i)}(u))_i) \\
&= \cup_{u \in A} \times_i g_i(\Pi_{R(i)}(u)) \\
&\subseteq \times_i g_i(\Pi_{R(i)}(A)).
\end{aligned}$$

Adding some assumptions, the following assertions are easy to prove.

Proposition 3.31. *Let $f, g \in \mathcal{R}(X)$ and $A \subseteq X$. Then*

$$f \cap g(A) = f(A) \cap g(A) \tag{3.11}$$
$$\text{if } \forall y \in Rg(f) \cap Rg(g), f^{-1}(y) \cap g^{-1}(y) \neq \emptyset,$$
$$f \backslash g(A) = f(A) \backslash g(A) \tag{3.12}$$
$$\text{if } \forall y \in Rg(f) \cap Rg(\sim g), f^{-1}(y) \cap (\sim g)^{-1}(y) \neq \emptyset,$$
$$f \times g(A) = f(\Pi_1(A)) \times g(\Pi_2(A)) \tag{3.13}$$
$$\text{if } A = \Pi_1(A) \times \Pi_2(A),$$
$$\otimes_R g_i(A) = \times_i g_i(\Pi_{R(i)}(A)) \tag{3.14}$$
$$\text{if } A = \times_i \Pi_{R(i)}(A).$$

Proof. Let us just prove (3.11), the next one follows the same argument and the last ones are trivial.

$$x \in f(A) \cap g(A)$$
$$\equiv \quad x \in f(A) \wedge x \in g(A)$$
$$\equiv \quad \exists u \in A, (u, x) \in f \wedge \exists u' \in A, (u', x) \in g$$
$$\because \quad f^{-1}(\{x\}) \cap g^{-1}(\{x\}) \neq \emptyset$$
$$\equiv \quad \exists u \in A, (u, x) \in f \wedge (u, x) \in g$$
$$\equiv \quad x \in f \cap g(A).$$

Remark 3.32. − Inversion is a basic operator. It cannot be replaced by something simpler, as in relation algebra.

− Among all items of Prop(s). 3.30 and 3.31, some of them can be seen as homomorphisms: Eq(s). (3.2), (3.6), (3.12), (3.11), (3.13), (3.14). Let us express the first one in its canonical form:

$$
\begin{array}{ccc}
f & \xrightarrow{\quad A \quad} & f(A) \\
{\scriptstyle(\ \leftarrow B)}\downarrow & & \downarrow {\scriptstyle\cap B} \\
(f \leftarrow B) & \xrightarrow{\quad A \quad} & (f \leftarrow B)(A) = f(A) \cap B
\end{array}
$$

The second one gives:

$$
\begin{array}{ccc}
f, g & \xrightarrow{\quad A \quad} & f(A), g(A) \\
{\scriptstyle\cup}\downarrow & & \downarrow {\scriptstyle\cup} \\
f \cup g & \xrightarrow{\quad A \quad} & f \cup g(A) = f(A) \cup g(A)
\end{array}
$$

Negation and difference are not easy to treat, as shown by Prop. 3.30. This is due to the fact that, (X, f) being a RDS and $A \subseteq X$, $\sim f(A)$ is generally not equal to $X \backslash f(A)$ but to $X \backslash \cap_{a \in A} f(a)$. We could define another unary operator providing such an "external" negation.

Definition 3.33 (External negation). *The external negation of a relation* $f \in \mathcal{R}(X, Y)$ *is given by*

$$\neg \quad : \quad \mathcal{R}(X, Y) \mapsto \mathcal{R}(X, Y)$$
$$s.t. \quad \forall A \subseteq X, \neg f(A) = X \backslash f(A).$$

We did not introduce this operator on equal footing with the other ones presented in §3.2 for two reasons: it is defined globally, using a set-transformer; the resulting set-transformer is not monotonic anymore, since set-difference is not either. In fact, $\neg f$ is anti-monotonic, but its square is again monotonic. Of course, both negations are related by the following: $\forall a \in X, A \subseteq X$,

$$\sim f(a) = \neg f(a)$$
$$\sim f(A) = X \backslash \cap_{a \in A} f(a) \supseteq X \backslash \cup_{a \in A} f(a) = \neg f(A).$$

In the following, we will try to go as far as possible using negation. However, in Chap. 6, we will examine external negation again.

The next two propositions are extensions of (3.6) and (3.7). To distinguish their names from Def. 2.37, we prefix their names with "meta-".

Proposition 3.34 (Universal meta-disjunctivity). *Let (X, f_i) be RDS, and $A \subseteq X$, then*

$$(\cup_n f_n)(A) = \cup_n f_n(A).$$

Remark 3.35. Without any restriction on the number and format of the composed relations f_i's, the resulting relation may not be closed. For example, let f_i be such that it maps its domain $[0, 1 - \frac{1}{i}]$ to a constant, say a. Every such relation is closed, but their infinite union is semi-closed: $Dom(\cup_i f_i) = [0, 1)$.

Proposition 3.36 (Semi meta-and-continuity). *Let (X, f_i) be RDS and $A \subseteq X$, then*

$$(\cap_n f_n)(A) \subseteq \cap_n f_n(A).$$

Remark 3.37. – This proposition simply extends Eq. (3.7). Since any intersection of closed sets is a closed set, the intersection $\cap_n f_n$ is always a closed relation on X.
– When "$\subseteq$" is replaced by "$=$", we get meta-and-continuity, which appears under some strong assumptions on relations. Our discussion of §2.3.3 concerning bounded non-determinism, image-finiteness and and-continuity also applies at the meta-level.

In §§2.2.2, 2.4 (Def. 2.76), we have presented discrete iteration schemes. It is easy to show that they can be equivalently rewritten as finite, infinite, transfinite sequential compositions. We have $f \in \mathcal{R}(X)$ and

$$f^0 = \mathcal{I},$$
$$f^{n+1} = f^n; f$$
$$= f; f^n.$$

Proposition 3.38. *Let f be a relation on X; and $A \subseteq X$. Then*

$$\forall n \geq 0, f^n(A) = \underbrace{f; \ldots; f}_{n \ times}(A).$$

Proof. Trivial by Def. 3.10, Prop. 3.30, and the general iteration scheme presented in Def. 2.76.

3.3.2 Point-Level Dynamics of Composed Systems

Here, we define the point-level dynamics of composed relations. In the following paragraphs, we review each case of composition individually, and we concentrate on the basic nondeterministic forward dynamics, i.e. $\theta(\cdot, \cdot)$.

Inversion. The unidirectional dynamics of the inverse of a relation f cannot be obtained easily using the dynamics of f itself. However, the complete bi-directional dynamics can be obtained by simple inversion of all sequences. If $s = \cdots s_{-2} s_{-1} s_0 s_1 s_2 \cdots \in X^{\mathbb{Z}}$, its mirror image is $\underline{s} = \cdots s_2 s_1 s_0 s_{-1} s_{-2} \cdots \in X^{\mathbb{Z}}$. Then, the dynamics of f is:

$$\Theta(A, f^{-1}) = \underline{\Theta(A, f)}.$$

Domain Restriction. Domain restriction is very easy to treat.

Proposition 3.39. *Let* (X, f) *be a RDS,* $A, B \subseteq X$, *then*

$$\theta(A, (B \to f)) = B^{\mathbb{N}} \cap \theta(A, f).$$

Proof.

$$
\begin{aligned}
& \theta(A, (B \to f)) \\
=\ & \{s \in X^{\mathbb{N}} \mid (s_0 \in A) \wedge (\forall n, (s_n, s_{n+1}) \in (B \to f))\} \\
=\ & \{s \in X^{\mathbb{N}} \mid (s_0 \in A) \wedge (\forall n, (s_n \in B) \wedge (s_n, s_{n+1}) \in f)\} \\
=\ & B^{\mathbb{N}} \cap \theta(A, f).
\end{aligned}
$$

Range restriction. Range restriction is trivial, too.

Proposition 3.40. *Let* (X, f) *be a RDS,* $A, B \subseteq X$, *then*

$$\theta(A, (f \leftarrow B)) = A \times B^{\mathbb{N}} \cap \theta(A, f).$$

Proof.

$$
\begin{aligned}
& \theta(A, (f \leftarrow B)) \\
=\ & \{s \in X^{\mathbb{N}} \mid (s_0 \in A) \wedge (\forall n, (s_n, s_{n+1}) \in (f \leftarrow B))\} \\
=\ & \{s \in X^{\mathbb{N}} \mid (s_0 \in A) \wedge (\forall n, (s_{n+1} \in B) \wedge (s_n, s_{n+1}) \in f)\} \\
=\ & A \times B^{\mathbb{N}} \cap \theta(A, f).
\end{aligned}
$$

Negation. To compute the dynamics of a negated system $\sim f$, it is important to remark that no sequence can contain a transition which belongs to f. This is stronger than removing $\theta(X, f)$ from the set of all possible sequences $X^{\mathbb{N}}$.

Proposition 3.41. *Let* (X, f) *be a RDS,* $A \subseteq X$, *then*

$$\theta(A, \sim f) = (A \times X^{\mathbb{N}}) \setminus \bigcup_{n=0}^{\infty} (X^n \times f \times X^{\mathbb{N}}).$$

Proof.

$$s \in \theta(A, \sim f)$$
$$\equiv (s \in X^{\mathbb{N}}) \wedge (s_0 \in A) \wedge (\forall n, (s_n, s_{n+1}) \notin f)$$
$$\equiv (s \in A \times X^{\mathbb{N}}) \wedge (\forall n, (s_n, s_{n+1}) \notin f)$$
$$\equiv (s \in A \times X^{\mathbb{N}}) \wedge (\forall n, s \notin (X^n \times f \times X^{\mathbb{N}}))$$
$$\equiv s \in (A \times X^{\mathbb{N}}) \setminus \bigcup_{n=0}^{\infty} (X^n \times f \times X^{\mathbb{N}}).$$

Sequential Composition. This case of composition is not very easy to study using information on individual components. Every sequence of $\theta(A, f; g)$ has the following form, with $s_0 \in A$:

$$s_0 \xrightarrow{f;g} s_1 \xrightarrow{f;g} s_2 \xrightarrow{f;g} \cdots$$

A "microscopic" view of this sequence gives

$$s_0 \xrightarrow{f} s_0' \xrightarrow{g} s_1 \xrightarrow{f} s_1' \xrightarrow{g} s_2 \xrightarrow{f} s_2' \xrightarrow{g} \cdots$$

where s_i' are intermediate states. This sequence can be obtained as the set of sequences whose head belongs to A, such that (s_0, s_0'), (s_1, s_1'), and every (s_i, s_i') belongs to f, and such that every $(s_i', s_{i+1}) \in g$. Then, a projection on unquoted indices is necessary to hide the details. Globally, this gives:

$$\theta(A, f; g) = \Pi_{\{2i | i \in \mathbb{N}\}}(A \times X^{\mathbb{N}} \cap f^{\mathbb{N}} \cap X \times g^{\mathbb{N}}).$$

Adding stronger assumptions on f, g or their composition could lead to nicer properties. We postpone these interesting developments to Chap. 6, where some stronger hypotheses will be used to compute the invariant of sequentially composed systems.

Intersection. The case of intersection is surprisingly simple. We have the following proposition.

Proposition 3.42. *Let (X, f) and (X, g) be RDS, $A \subseteq X$, then*

$$\theta(A, f \cap g) = \theta(A, f) \cap \theta(A, g).$$

Proof.

$$\theta(A, f \cap g)$$
$$= \{s \in X^{\mathbb{N}} \mid (s_0 \in A) \wedge (\forall n, (s_n, s_{n+1}) \in f \wedge (s_n, s_{n+1}) \in g)\}$$
$$= \{s \in A \times X^{\mathbb{N}} \mid (\forall n, (s_n, s_{n+1}) \in f) \wedge (\forall n, (s_n, s_{n+1}) \in g)\}$$
$$= \theta(A, f) \cap \theta(A, g).$$

Remark 3.43. In trace-based parallelism semantics, the equivalent statement holds. If S_1 and S_2 represent programs or processes, their synchronized parallel composition $S_1 \parallel S_2$ is such that

$$[[S_1 \parallel S_2]] = [[S_1]] \cap [[S_2]]$$

where $[[S]]$ denotes the semantics of S, i.e. its set of possible traces or trajectories [79].

Union. This composition is much more complicated than intersection. To get the complete dynamics of the union of two systems f and g, we have to take into account all sequences obtained by arbitrary sequential compositions of f and g. Let us start with a simple motivating example.

Example 3.44. Let us imagine that we have two systems f and g defined on the same space X, with the same domain and range, and that we want to compose them by a nondeterministic choice at each iteration step. In our framework, this is possible using the union of relations. The resulting dynamics is much richer than the union of the individual sets of evolutions. The global dynamics contains all interleaved execution traces

For instance, $X = \{a, b\}$, $f = \{(a,a),(b,b)\}$ and $g = \{(a,b),(b,a)\}$. The corresponding dynamics are

$$\begin{aligned}
\theta(X, f) &= \{a \longrightarrow a \longrightarrow a \longrightarrow \cdots, \\
&\quad\ b \longrightarrow b \longrightarrow b \longrightarrow \cdots\} \\
\theta(X, g) &= \{a \longrightarrow b \longrightarrow a \longrightarrow \cdots, \\
&\quad\ b \longrightarrow a \longrightarrow b \longrightarrow \cdots\}.
\end{aligned}$$

The union of individual dynamics is much simpler:

$$\begin{aligned}
\theta(X, f) \cup \theta(X, g) &= \{a \longrightarrow a \longrightarrow a \longrightarrow \cdots, \\
&\quad\ b \longrightarrow b \longrightarrow b \longrightarrow \cdots, \\
&\quad\ a \longrightarrow b \longrightarrow a \longrightarrow \cdots, \\
&\quad\ b \longrightarrow a \longrightarrow b \longrightarrow \cdots\}.
\end{aligned}$$

The composite relation is $f \cup g = \{(a,a),(a,b),(b,a),(b,b)\}$, which generates a more "complex" dynamics:

$$\theta(X, f \cup g) = \{a, b\}^\omega.$$

Actually, we can approximate this set of sequences from below by adding larger and larger periodic sequential compositions: first, we consider f (giving rise to $f; f; f; \cdots$) and g (leading to $g; g; g; \cdots$); second, we consider 2-periodic repetitions of $f; f$, $f; g$, $g; f$ and $g; g$; and so on, and so forth.

Thus, we have to generalize our previous presentation of sequential composition. Therefore, we introduce a modified point-level dynamics that realizes a microscopic view of sequential composition:

$$\theta_\mu(A, (f_j)_{j=1}^k) = \{s \in X^\mathbb{N} \mid (s_0 \in A) \land (\forall n, (s_n, s_{n+1}) \in f_{(n \bmod k)+1})\}.$$

Let us denote sequential compositions as words, e.g. fgg instead of $f; g; g$. In principle, we have

$$\theta(A, f \cup g) = \cup_{w \in \{f,g\}^\mathbb{N}} \theta_\mu(A, w)$$

but we can also compute the closure of the approximations since the set of periodic infinite words P is dense in the set of infinite words $\{f, g\}^{\mathbb{N}}$, whence $\overline{P} = \{f, g\}^{\mathbb{N}}$ [96, 328].

Proposition 3.45. Let (X, f) and (X, g) be RDS, and $A \subseteq X$, then

$$\theta(A, f \cup g) = \overline{\cup_{w \in \{f,g\}^*} \theta_\mu(A, w)}.$$

Proof. We have to prove that $\forall s \in \theta(A, f \cup g), \forall \varepsilon, \exists t \in \cup_{w \in \{f,g\}^*} \theta_\mu(A, w), d_a(s, t) < \varepsilon$.

Let us fix ε and take the smallest k such that $2^{-k} \le \varepsilon$. Then, t such that $t|_k = s|_k$ is sufficient. It remains to show that this t belongs to the union: as $s \in \theta(A, f \cup g)$, there exists a word $w \in \{f, g\}^\omega$ such that $\forall i, (s_i, s_{i+1}) \in w_i$ and we have of course $w|_k \in \{f, g\}^k$ and $t \in \theta_\mu(A, w|_k)$.

Difference. This composition follows the same lines as negation. The next proposition states the result.

Proposition 3.46. Let (X, f) and (X, g) be RDS, $A \subseteq X$, then

$$\theta(A, f \backslash g) = \theta(A, f) \backslash \cup_{n=0}^\infty (X^n \times g \times X^{\mathbb{N}}).$$

Proof.

$$
\begin{aligned}
&\theta(A, f \backslash g) \\
=\ & \{s \in X^{\mathbb{N}} \mid (s_0 \in A) \wedge (\forall n, (s_n, s_{n+1}) \in f \wedge (s_n, s_{n+1}) \notin g)\} \\
=\ & \theta(A, f) \cap \theta(A, \sim g) \\
=\ & \theta(A, f) \cap (A \times X^{\mathbb{N}} \backslash \cup_n (X^n \times g \times X^{\mathbb{N}})) \\
=\ & \theta(A, f) \backslash \cup_n (X^n \times g \times X^{\mathbb{N}}).
\end{aligned}
$$

Free Product. Now, we analyze the product of systems without interactions between components. Before giving the next proposition, let us define a specific projection: let s be a sequence of $(X \times Y)^{\mathbb{N}}$, then $\Pi^1(s)$ is obtained by extraction of the first components:

$$\Pi^1((s_0^1, s_0^2)(s_1^1, s_1^2)(s_2^1, s_2^2) \cdots) = s_0^1 s_1^1 s_2^1 \cdots$$

and $\Pi^2(s)$ applies symmetrically on second components.

Proposition 3.47. Let (X, f) and (Y, g) be RDS, $A \subseteq X$, $B \subseteq Y$, then

$$\theta(A \times B, f \times g) = \{s \in (X \times Y)^{\mathbb{N}} \mid (\Pi^1(s) \in \theta(A, f)) \wedge (\Pi^2(s) \in \theta(B, g))\}.$$

Proof. Trivial using Def. 2.9 and the projectors Π^i.

Remark 3.48. We can rewrite this expression in a simple manner, if we relax the ordering imposed by Cartesian products, i.e. $X^{\mathbb{N}} \times Y^{\mathbb{N}} =_\Pi (X \times Y)^{\mathbb{N}}$:

$$\theta(A \times B, f \times g) =_\Pi \theta(A, f) \times \theta(B, g).$$

Connected Product. Finally, the case of connected products is very difficult to treat in general. Indeed, the sequence of pairs of states cannot be decomposed as we did it in the previous case (for free products), which eliminates many potential simplifications. Thus, the complete dynamics of this last composition will not be analyzed further on.

Summary of Compositional Results.. Some operators permit a compositional analysis of their point-level dynamics: inversion, restrictions, intersection, free product. Strictly speaking, the other operators are not compositional. These properties are summarized in Table 3.1, where (X, f) and (X, g) are RDS, and $A, B \subseteq X$.

Table 3.1. Compositional analysis of point-level dynamics

Compositional results		
$\Theta(A, f^{-1})$	$=$	$\Theta(A, f)$
$\theta(A, (B \to f))$	$=$	$\overline{B^{\mathbb{N}} \cap \theta(A, f)}$
$\theta(A, (f \leftarrow B))$	$=$	$A \times B^{\mathbb{N}} \cap \theta(A, f)$
$\theta(A, f \cap g)$	$=$	$\theta(A, f) \cap \theta(A, g)$
$\theta(A \times B, f \times g)$	$=_\Pi$	$\theta(A, f) \times \theta(B, g)$
Non-compositional results		
$\theta(A, f \cup g)$	$=$	$\overline{\cup_{w \in \{f,g\}^*} \theta_\mu(A, w)}$
$\theta(A, f \backslash g)$	$=$	$\theta(A, f) \backslash \cup_{n=0}^{\infty} (X^n \times g \times X^{\mathbb{N}})$
$\theta(A, \sim f)$	$=$	$(A \times X^{\mathbb{N}}) \backslash \cup_{n=0}^{\infty} (X^n \times f \times X^{\mathbb{N}})$
$\theta(A, f; g)$	$=$	$\Pi_{\{2i \mid i \in \mathbb{N}\}} (A \times X^{\mathbb{N}} \cap f^{\mathbb{N}} \cap X \times g^{\mathbb{N}})$

3.4 Algebraic Properties of Composition Operators

In §3.2, we have defined composition operators to combine relations into new ones. These operators can in turn be combined to each other, to form more and more complex systems (see Def. 3.25). To facilitate the technical developments based on these operators, it is interesting to study algebraic properties of operators. This section aims at studying the composition of composition operators.

The operators on relations presented above can have properties like associativity, commutativity, distributivity on each others, etc. The proof of these properties is straightforward, therefore we do not detail them and refer the interested reader to [245] for example, to get an idea of some of them, or to any textbook on set theory, e.g. [141].

At first, let us examine some $(n > 1)$-ary operators. It is clear that:

- ; is associative;
- $\cup$ and $\cap$ are idempotent, associative and commutative.

The way we have defined $\times$ and $\otimes$, these operators are neither associative, nor commutative, because we work with tuples of tuples instead of working with flat vectors.

We can compose all these operators together, which leads to some interesting properties. We first compose unary operators, then mix unary and $(n > 1)$-ary operators, and we finally show some compositions of $(n > 1)$-ary operators.

To represent the composition of operators, we use the symbol $\circ$: if $\star$ and $\star'$ are two operators, $\star \circ \star'$ represents the successive application of $\star'$ and $\star$. For instance, let r be a relation, $\star$ and $\star'$ be two unary operators,

$$\star \circ \star'(r) = \star(\star'(r)).$$

We also use the "curryfication" of binary operations in the following way: $\cap$ can be seen as a binary operator taking two arguments, A and B, which gives $A \cap B$, but it is also possible to fix one of these arguments, say A, which gives a unary operator $\cap A$ taking one argument, B, which again gives $A \cap B$.

The composition $\circ$ has an identity, denoted by id. This means that it is equivalent to the application of the identity relation.

Remark 3.49. – In the following tables, the relations to which these compositions of composition operators are supposed to be applied are defined on a unique space X. Generalizing this to any other space is straightforward.
- Since the connected product is based on projections, a simplification of its composition with other operators cannot be systematically obtained. This is the case for the composition with unary operators, or with \ and $\times$. Thus, we express the composition whenever it leads to a simpler form.

3.4.1 Composition of Unary Operators

The composition of unary operators is given in Table 3.2. Each cell (row $\star$, col. $\star'$) of the tables gives a composition of two unary operators, $\star' \circ \star$, e.g. $^{-1} \circ \sim = \sim \circ ^{-1}$.

3.4.2 Composition of Unary and N-Ary Operators

The composition of unary and $(n > 1)$-ary operators is given in Table 3.3. The application of two unary operators $\star$ and $\star'$ to two relations is denoted by $(\star, \star')$; $(\star)^2$ stands for $(\star, \star)$.

Table 3.2. Composition of unary operators

$\circ$	$^{-1}$	$\sim$
$^{-1}$	$\mathcal{I}$	$^{-1}\circ\sim$
$(B\to)$	$(\leftarrow B)\circ\,^{-1}$	$\sim\circ\cap(B\times X)$
$(\leftarrow B)$	$(B\to)\circ\,^{-1}$	$\sim\circ\cap(X\times B)$
$\sim$	$\sim\circ\,^{-1}$	$\mathcal{I}$

$\circ$	$(A\to)$	$(\leftarrow A)$
$^{-1}$	$^{-1}\circ(\leftarrow A)$	$^{-1}\circ(A\to)$
$(B\to)$	$((A\cap B)\to)$	$(B\to)\circ(\leftarrow A)$
$(\leftarrow B)$	$(\leftarrow B)\circ(A\to)$	$(\leftarrow(A\cap B))$
$\sim$	$\cap(A\times X)\circ\sim$	$\cap(X\times A)\circ\sim$

3.4.3 Composition of N-Ary Operators

Finally, we investigate the composition of $(n>1)$-ary operators. Since it requires more attention, we split the analysis in several cases.

Table 3.3. Composition of unary and $(n>1)$-ary operators

$\circ$	$(B\to)$	$(\leftarrow B)$
;	$;\circ((B\to),id)$	$;\circ(id,(B\to))$
$\cap$	$\cap\circ((B\to))^2$	$\cap\circ((\leftarrow B))^2$
$\cup$	$\cup\circ((B\to))^2$	$\cup\circ((\leftarrow B))^2$
$\backslash$	$\backslash\circ((B\to),id)$	$\backslash\circ((\leftarrow B),id)$
$\times$	$\times\circ((B_1\to),(B_2\to))$	$\times\circ((\leftarrow B_1),(\leftarrow B_2))$
	if $\exists B_1,B_2$ s.t. $B=B_1\times B_2$	

$\circ$	$^{-1}$	$\sim$
;	$;\circ(^{-1})^2$	$\cup\circ(\times X\circ\Pi_1\circ\sim,X\times\circ\Pi_1\circ\sim)$
$\cap$	$\cap\circ(^{-1})^2$	$\cup\circ(\sim)^2$
$\cup$	$\cup\circ(^{-1})^2$	$\cap\circ(\sim)^2$
$\backslash$	$\backslash\circ(^{-1})^2$	$\cup\circ(\sim,id)$
$\times$	$\times\circ(^{-1})^2$	$\supseteq\times\circ(\sim)^2$

Let us begin with commutative operators, $\cup$ and $\cap$ (see Table 3.4).

Remark 3.50. The last line of the previous table holds because Π_R is simple (we use the second line of the following table).

In Table 3.5, composition with ;.
In Table 3.6, composition with $\backslash$.
Finally, in Table 3.7, composition with $\times$.

Table 3.4. Composition of $(n > 1)$-ary operators with commutative operators

∘	$\cap h$	$\cup h$
$f \cap g$	$f \cap (g \cap h)$	$(f \cup h) \cap (g \cup h)$
$f \cup g$	$(f \cap h) \cup (g \cap h)$	$f \cup (g \cup h)$
$f \backslash g$	$(f \cap h) \backslash g$	$(f \cup h) \cap (\sim g \cup h)$
$f \times g$	$(f \cap h_1) \times (g \cap h_2)$	$(f \cup h_1) \times (g \cup h_2)$
	if $\exists h_1, h_2$ s.t. $h = h_1 \times h_2$	
$f \otimes_R g$	$(f \cap h_1) \otimes_R (g \cap h_2)$	$(f \cup h_1) \otimes_R (g \cup h_2)$
	if $\exists h_1, h_2$ s.t. $h = h_1 \otimes_R h_2$	

Table 3.5. Composition of $(n > 1)$-ary operators with ;

∘	$; h$	$h;$
$f ; g$	$f ; (g ; h)$	$(h ; f) ; g$
$f \cap g$	$(f ; h) \cap (g ; h)$	$(h ; f) \cap (h ; g)$
	if h injective	if h simple
$f \cup g$	$(f ; h) \cup (g ; h)$	$(h ; f) \cup (h ; g)$
$f \backslash g$	$(f ; h) \cap (\sim g ; h)$	$(h ; f) \cap (h ; \sim g)$
	if h injective	if h simple
$f \times g$	$(f ; h_1) \times (g ; h_2)$	$(h_1 ; f) \times (h_2 ; g)$
	if $\exists h_1, h_2$ s.t. $h = h_1 \times h_2$	
$f \otimes_R g$	/	$(f ; h_1) \otimes_R (g ; h_2)$
	if $\exists h_1, h_2$ s.t. $h = h_1 \times h_2$	

Table 3.6. Composition of $(n > 1)$-ary operators with $\sim$

∘	$\backslash h$	$h \backslash$
$f \cap g$	$(f \backslash h) \cap (g \backslash h)$	$(h \backslash f) \cup (h \backslash g)$
$f \cup g$	$(f \backslash h) \cup (g \backslash h)$	$(h \backslash f) \cap (h \backslash g)$
$f \backslash g$	$f \backslash (g \cup h)$	$(h \backslash f) \cup (h \cap g)$
$f \times g$	$(f \backslash h_1) \times (g \backslash h_2)$	$(h_1 \backslash f) \times (h_2 \backslash g)$
	if $\exists h_1, h_2$ s.t. $h = h_1 \times h_2$	

Table 3.7. Composition of $(n > 1)$-ary operators with $\times$

∘	$\times h$	$h \times$
$f \cap g$	$(f \times h) \cap (g \times h)$	$(h \times f) \cap (h \times g)$
$f \cup g$	$(f \times h) \cup (g \times h)$	$(h \times f) \cup (h \times g)$
$f \backslash g$	$(f \times h) \backslash (g \times h)$	$(h \times f) \backslash (h \times g)$

3.4.4 Fixpoint Theory for the Composition

So far, we have defined relational dynamical systems (Def. 2.6), and composed dynamical systems (Def. 3.25) based on composition operators used to organize the structure of systems explicitly.

Here we generalize this construction by introducing an important family of structured systems: recursive relations defined as fixpoints of *functional equations*. Let us illustrate these by an example.

Example 3.51. We consider a functional equation defined as follows: f_1, and f_2 being RDS defined on X,

$$\Gamma(f) = f_1 \cup (f_2; f).$$

We would like to define r as the greatest fixpoint of Γ, i.e. $\mathbb{G}\Gamma$. How to do that?

In Chap. 2, we have presented several lattice fixpoint theorems, including a transfinite version requiring monotonic functions only (§§2.3.4, 2.4). These theorems are general enough to support the meta-level we need here, because relations are nothing but elements of the power set of a Cartesian space, which is always a complete lattice.

The assumptions required on functions that are used to define fixpoint equations are monotonicity and, restricting ourselves to ω iterations, and-continuity and or-continuity.

Thus, we first have to define an ordering on relations. Second, we must exhibit functionals that preserve this order (monotonicity) and that preserve limits (continuity). We concentrate on monotonicity, for we have seen that continuity can be replaced by transfinite iteration.

Many composition operators preserve the order, even applied to infinite monotonic sequences of relations. We define a least upper bound of some infinite sequences, and a notion of fixpoint (in this case, fixed relation) of functional meta-equations. Using a notion of continuity, this fixpoint can be computed by successive applications of the functional, starting from an empty relation, an undefined relation (least fixpoint), or from a total relation (greatest fixpoint).

Ordering Relations. It is possible to define several orders on relations. For example, in [245], two partial orders on $\mathcal{R}(X)$ are considered: the *inclusion order*,

$$(f \subseteq g) \Leftrightarrow (\forall u, f(u) \subseteq g(u))$$

and the *definition order*,

$$(f \sqsubseteq g) \Leftrightarrow (\forall u, f(u) = \emptyset \vee f(u) = g(u)).$$

It is easy to verify that

$$(f \sqsubseteq g) \Rightarrow (f \subseteq g).$$

Order-Preserving Functionals. The functionals we need here are based on composition operators we have presented in §3.2. Let us review some of them w.r.t. monotonicity. Since we just rephrase standard properties on relations, we list the results without proofs (see e.g. [245, §1.3.3] and [246]).

Proposition 3.52 ($\subseteq$-meta-monotonicity). *If f, g, h are RDS defined on X, such that $f \subseteq g$, and $\forall A \subseteq X$, then the following statements hold:*

$$
\begin{aligned}
(A \rightarrow f) &\subseteq (A \rightarrow g) \\
(f \leftarrow A) &\subseteq (g \leftarrow A) \\
f; h &\subseteq g; h \\
h; f &\subseteq h; g \\
f \cup h &\subseteq g \cup h \\
f \cap h &\subseteq g \cap h \\
f \times h &\subseteq g \times h.
\end{aligned}
$$

Proposition 3.53 ($\sqsubseteq$-meta-monotonicity). *If f, g, h are RDS defined on X, such that $f \sqsubseteq g$, and $\forall A \subseteq X$, then the following statements hold:*

$$
\begin{aligned}
(A \rightarrow f) &\sqsubseteq (A \rightarrow g) \\
(f \leftarrow A) &\sqsubseteq (g \leftarrow A) \\
h; f &\sqsubseteq h; g \\
f \cup h &\sqsubseteq g \cup h \\
&\quad \textit{if } Dom(h) \cap (Dom(f) \cup Dom(g)) = \emptyset \\
f \cap h &\sqsubseteq g \cap h \\
f \times h &\sqsubseteq g \times h.
\end{aligned}
$$

Fixpoint Meta-Equations. Using fixpoint theorems of §2.3.4, we know that the greatest fixpoint of a monotonic, and-continuous functional Γ can be obtained as follows:

$$\mathbb{G}\Gamma = \cap_i \Gamma^i(\mathcal{U})$$

and its least fixpoint, in case of or-continuity, as

$$\mathbb{S}\Gamma = \cup_i \Gamma^i(\mathcal{B})$$

where $\mathcal{B}$ is a bottom relation like $\mathcal{E}$ or a more specific "undefined" relation.

Let us close the section with an illustration of the method.

Example 3.54 (Ex. 3.51 revisited). To compute the least fixpoint of Γ defined by $\Gamma(f) = f_1 \cup (f_2; f)$, we use the successive approximations from the empty relation $\mathcal{E}$:

$$\mathcal{E} \longrightarrow f_1 \cup f_2; \mathcal{E} = f_1 \longrightarrow f_1 \cup f_2; f_1 \longrightarrow f_1 \cup f_2; (f_1 \cup f_2; f_1) \longrightarrow \cdots$$

This sequence has a well-defined limit thanks to monotonicity of the involved operators $\cup$ and $;$.

3.5 Discussion

In this chapter, we introduced composition operators for relational discrete-time dynamical systems, and developed a compositional approach for their analysis. The approach is inspired by compositional ideas present in computer science and logic, more specifically in program theory and in parallelism semantics. Surprisingly, only a few attempts to apply structural composition to dynamical systems were proposed before [290, 244]. Here, the compositional approach holds for operations such as inversion, restrictions, intersection, free product but not for the other operators like union and connected product.

In §3.3.1, we analyzed set-transformers of composed relation (e.g. see Prop. 3.30). Equivalent results on predicate-transformers and multi-valued functions can be found in [245, 246].

In §3.3.2, the point-level dynamics of composed systems was studied (see Table 3.1). Similar results can be found in trace-based semantics of sequential and parallel programs and processes (e.g. [151, 79, 89]).

Here, we discuss related notions and results about composition operators, nondeterminism and probabilities, and fixpoint operators as composition means.

3.5.1 Composition Operators

The compositional idea is present in many works in computer science [94, 69, 329, 179, 299, 1, 67]. Our operators are inspired from different sources: logic, relation and graph theory, program theory, models of parallelism, and classical algebra [91, 245, 246, 53, 277].

In particular, domain and range restriction are sometimes called left and right restriction [245], sequential composition is called composition or multiplication [277], union is called sum [290] and used as composition means in [328, 159, 140, 325]. Free product is called Cartesian product, and the connected product is close to the relational notion of natural join [277, 53]. Our connected product is original in the sense we express the structure of cellular automata, neural networks, and other distributed systems in a uniform abstract way. Moreover, it can be particularized to well-known operators like synchronization products [19, 288].

Some models of concurrency, like transition systems for instance, include composition operators [19, 151, 222]. They serve as descriptive or algebraic elements for the study of complex systems built from simple ones.

The set of operators presented above is not minimum. Our goal is to have simple notations and simple concepts, which is not always possible when too few operators are defined. We try not to tend to the opposite extreme where too many operators give unclear frameworks.

Thanks to the operators we have presented and the iteration schemes we use, we are able to model synchronous (products) and asynchronous (union and nondeterminism, see §3.5.2) parallelism. Iterations are clocked but "silent" transitions can be added (identity relation). Thus, asynchronous evolution is also permitted. Different synchronization of processes can be expressed using our connected product, and specific projections.

3.5.2 Nondeterminism and Probabilities Revisited

This section further discusses nondeterminism and probabilities (see §2.5.4). Along the same line, it can be useful to introduce a choice, probabilistic or nondeterministic, between several components of a union. Notice that in this case, the choice is made globally, i.e. on the component, and not locally, i.e. not on the image by a component. This allows us to propose several variants of the union presented before (see Def. 3.15):

− a *nondeterministic union*, $f \cup_N g(x)$ is equal to

$$
\begin{array}{lll}
f(x) & \text{if} & x \in Dom(f) \backslash Dom(g) \\
g(x) & \text{if} & x \in Dom(g) \backslash Dom(f) \\
\emptyset & \text{if} & x \notin Dom(f) \cup Dom(g) \\
(!(\{f, g\}))(x) & \text{if} & x \in Dom(f) \cap Dom(g);
\end{array}
$$

Evolutions based on this multi-valued function are between $\theta(X, f \cup g)$ and $\xi(X, f \cup g)$ because they sometimes take a part of the full answer (last line), but all nondeterministic evolutions of $f \cup g$ are present in $\theta(X, f \cup_N g)$.
− a *fair nondeterministic union*, based on the same definition as the previous one, but adding an assumption of fairness over all infinite histories of the system, i.e. no component can be forgotten forever;
In this case, the complete dynamics is modified because all infinite sequential compositions terminating with an infinite composition of f or g must be removed (provided that f and g are not equal):

$$
\theta(X, f \cup_F g) = \theta(X, f \cup g) \backslash \cup_n ((X^n \times \theta(X, f)) \cup (X^n \times \theta(X, g))).
$$

The dynamics of this fair nondeterministic union is called "pseudo-periodic chaotic iteration" in [274].
− a *probabilistic union*, based on a vector $p = (p_i)_i$ of probabilities, i.e. $\forall i, p_i \geq 0$ and $\sum_i p_i = 1$, $f \cup_{P:p} g(x)$ is equal to

$$
\begin{array}{lll}
f(x) & \text{if} & x \in Dom(f) \backslash Dom(g) \\
g(x) & \text{if} & x \in Dom(g) \backslash Dom(f) \\
\emptyset & \text{if} & x \notin Dom(f) \cup Dom(g) \\
p_1 : f(x) & \text{if} & x \in Dom(f) \cap Dom(g) \\
p_2 : g(x) & &
\end{array}
$$

which means that when both branches are allowed, f is chosen with probability p_1, and g with probability p_2.

Theoretically, these operators are not very different from the union but their use can speed up iterations from a simulation point of view.

The "wholistic" interpretation of these variants of union in terms of set-transformers are all equivalent to the original one: $\forall A \subseteq X$,

$$(f \cup_N g)(A) = (f \cup_F g)(A) = (f \cup_{P:p} g)(A) = (f \cup g)(A)$$

but the dynamics can be modified, as we have seen it in case of fairness.

Of course, we wonder about the meaning of having a deterministic set-transformer evolution whereas particular evolutions are based on probabilities for example. In fact, instead of working with sets, where inclusion is binary ($a \in A$ or $a \notin A$), a first generalization consists in adding weights to states (e.g. $w(a, A) = 0.145$), thereby expressing a degree of inclusion. This leads to measures. A second generalization consists in generating measures at random, on top of probability densities.

These two generalizations are explored in fractal theory [160]. The first one has also been introduced in program semantics, by extension of predicate-transformers to probabilistic program executions [234]. It could be interesting to introduce them at the level of our set-transformers.

3.5.3 Fixpoint Operator and Composition

We have introduced iteration as external operation on relations, which are potentially composed objects. However, in §3.4.4, we have added a possibility of defining relations as fixpoint solutions of functional equations, using $\mathbb{S}$ and $\mathbb{G}$, as proposed in [245] for relations expressing the semantics of programs.

For instance, infinite iteration can be seen as infinite sequential composition, which in turn can be seen as the greatest fixpoint solution of a specific functional equation. Thus, f being a relation, infinite composition f^ω can be seen as a composition operator. Then, we can mix it with the other operators we have defined in this chapter. For instance, an expression like

$$(f^\omega \cup (g; h^\omega))^\omega$$

can be meaningful, even if it takes a long time before reaching a final state.

An interesting extension of our framework would be the integration of fix-points of functional equations as composition operators. In short, the questions are: How do set-transformers and point-level dynamics behave w.r.t. fixpoint operators $\mathbb{S}$ and $\mathbb{G}$? How are they composed with other composition operators?

4. Abstract Observation of Dynamics

In Chap(s). 2 and 3, we defined composed relational dynamical systems, and we presented two equivalent ways to look at the evolution of systems: a point-level nondeterministic dynamics, and a set-level deterministic dynamics. In this chapter, we introduce the observation of dynamical systems, and their abstraction.

The fully precise observation of evolutions of dynamical systems is not always possible: infinite precision is beyond human capability, and a coarse-grained observation of evolution states is often necessary or simply more realistic.

Trajectories are effective sequences of states. In the observation of systems, we focus on transitions between states, which leads to the notion of trace. Set-transformers and nondeterministic dynamics can be parametrized by observed evolutions of systems.

Effective and observed evolutions of dynamical systems can be abstracted, viz. simplified: states or groups of states are replaced by abstract states, and the concrete dynamical system is replaced by an abstract one. Under some conditions, some qualitative properties on the dynamics, e.g. like invariance and reachability, can be proved at the abstract level and remain valid at the concrete level.

This short chapter is organized as follows: we first introduce observation traces in §4.1, and traced-based evolutions in §4.2; we particularize observations to symbolic traces in §4.3; abstraction is presented in §4.4, and we show in §4.5 that qualitative dynamical properties can be preserved by abstraction; in §4.6, we consider observation as an abstraction homomorphism; finally, we close the chapter in §4.7 with a discussion.

4.1 Observation of Systems

The dynamics of a system can be defined by its set of possible trajectories (see Def. 2.9). A trajectory is an ordered sequence of states visited during a particular evolution of the system. A dual view of trajectories consists in looking at transitions between states. The observation of such a transition sequence is a trace.

F. Geurts: Abstract Compositional Analysis of Iterated Relations, LNCS 1426, pp. 83-94, 1998.
© Springer-Verlag Berlin Heidelberg 1998

Let us introduce observation functions: they establish a correspondence between transitions states and labels. When a fully precise observation of transitions is possible, the correspondence is bijective; in general, it is surjective.

Definition 4.1 (Observation function). *An observation function of a RDS (X, f) is a surjective function $\mathcal{V} : X \times X \mapsto \mathcal{X}$, totally defined from state transitions of f to a compact metric space $\mathcal{X}$.*

Remark 4.2. − The function must be total on f but not necessarily on all of $X \times X$.
− The compact metric space $\mathcal{X}$ is called *label space* because its elements label state transitions.

Any admissible state transition of f is associated with a label given by the observation function.

Definition 4.3 (Observed transition). *Given a RDS (X, f) and an observation function $\mathcal{V} : X \times X \mapsto \mathcal{X}$, an observed transition of f is a label $e \in \mathcal{X}$ such that $\exists x_1, x_2 \in X, \mathcal{V}(x_1, x_2) = e$.*

Definition 4.4 (Trace). *Given a label space $\mathcal{X}$, a trace is any word on $\mathcal{X}$.*

Not all traces are possible observations of a given dynamical system. An observed traced is an admissible sequence of label transitions.

Definition 4.5 (Observed trace). *Given a RDS (X, f) and an observation function $\mathcal{V} : X \times X \mapsto \mathcal{X}$, a trace of f is a sequence $\sigma \in \mathcal{X}^\infty$ such that $\exists s \in \theta(X, f), \forall n, \mathcal{V}(s_n, s_{n+1}) = \sigma_n$.*

Notation 4.6. *For simplicity, we extend the observation function from state transitions to trajectories. Thus, in the previous definition, $\mathcal{V}(s) = \sigma$.*

Remark 4.7. An "interesting" observation function is a trade-off: on one hand, full precision is often difficult to get; on the other hand, the coarse-graining must be sufficiently precise to be useful, which means that a large amount of traces must be observable. This will be discussed later on in this chapter, and in Chap. 5, where infinite traces will be related to specific invariant states through fullness and atomicity properties.

In general, we can define a set containing all observed traces.

Definition 4.8 (Trace language). *Given a RDS (X, f) and an observation function $\mathcal{V} : X \times X \mapsto \mathcal{X}$, the corresponding trace language $L_t(X, f, \mathcal{X}, \mathcal{V})$ is the set of all observed traces.*

Remark 4.9. In the following, we will sometimes need bidirectional traces and corresponding trace languages; these can be easily defined from Def. 2.8 and Rem. 2.11. However, for simplicty, we restrict the presentation of this chapter to unidirectional traces.

4.2 Trace-Based Dynamics

Given an observation function of the RDS (X, f), both set-level and point-level dynamics can be related to particular transitions and traces. In order to do that, f must first be parametrized by transitions.

Definition 4.10 (Transition-parametrized set-transformer). *Let* (X, f) *be a RDS, and* $\mathcal{V} : X \times X \mapsto \mathcal{X}$ *an observation function. Any transition* $e \in \mathcal{X}$ *induces a parametrized set-transformer*

$$f_e = \{(x, y) \in f \mid \mathcal{V}(x, y) = e\}.$$

The following proposition is interesting in that it allows to decompose f into transition-parametrized components.

Proposition 4.11. *Let* (X, f) *be a RDS, and* $\mathcal{V} : X \times X \mapsto \mathcal{X}$ *an observation function. Then,*

$$f = \cup_{e \in \mathcal{X}} f_e.$$

Proof. Since $\mathcal{V}$ is total on f and surjective onto $\mathcal{X}$, we have

$$f = \{(x, y) \in f \mid \exists e \in \mathcal{X}, \mathcal{V}(x, y) = e\}.$$

Now, set-transformers can be parametrized by traces.

Definition 4.12 (Trace-parametrized set-transformer). *Let* (X, f) *be a RDS, and* $\mathcal{V} : X \times X \mapsto \mathcal{X}$ *an observation function. Any trace* $\sigma \in \mathcal{X}^\infty$ *induces a parametrized set-transformer* f_σ: $\forall e \in \mathcal{X}$,

$$
\begin{aligned}
f_\varepsilon &= \mathcal{I}_X \\
f_{e\sigma} &= f_e; f_\sigma \\
f_{\sigma e} &= f_\sigma; f_e.
\end{aligned}
$$

Notation 4.13 (Representation of histories). *In case of an inverse relation* f^{-1}, $e\sigma$ *represents a past history. Then,* e *is close to present and* σ *describes the subsequent backward evolution.*

Every nonempty trace σ *on* $\mathcal{X}$, *finite of infinite, will always be represented from left to right. However, sometimes, it will be useful to consider inverse histories, in which case we will use a right-juxtaposed comma to denote the backward representation. Thus, the following notations will be useful:* $\forall \sigma, \tau \in \mathcal{X}^\infty$,

$$
\begin{aligned}
\sigma &= \sigma_0 \sigma_1 \cdots \\
, \sigma &= \sigma_0 \sigma_1 \cdots \\
\sigma, &= \cdots \sigma_1 \sigma_0 \\
\sigma, \tau &= \cdots \sigma_1 \sigma_0, \tau_0 \tau_1 \cdots
\end{aligned}
$$

The comma indicates where the trace starts. If both σ and τ are infinite traces, then (σ, τ) is a bi-infinite trace.

Finally, the point-level dynamics can also be parametrized by traces.

Definition 4.14 (Trace-parametrized point-level dynamics). *Let (X, f) be a RDS, and $\mathcal{V} : X \times X \mapsto \mathcal{X}$ an observation function. Then, $\forall A \subseteq X$, and $\forall \sigma \in \mathcal{X}^{\mathbb{N}}$,*

$$\theta_\sigma(A, f) = \{s \in X^{\mathbb{N}} \mid (s_0 \in A) \wedge (\forall n, (s_n, s_{n+1}) \in f_{\sigma_n})\}.$$

Of course, the full dynamics can be decomposed in the same way as the complete set-transformer.

Proposition 4.15. *Let (X, f) be a RDS, and $\mathcal{V} : X \times X \mapsto \mathcal{X}$ an observation function. Then, $\forall A \subseteq X$,*

$$\theta(A, f) = \cup_{\sigma \in \mathcal{X}^{\mathbb{N}}} \theta_\sigma(A, f).$$

Proof.

$$
\begin{aligned}
& s \in \theta(A, f) \\
\equiv\ & (s_0 \in A) \wedge \forall n, (s_n, s_{n+1}) \in f \\
\equiv\ & (s_0 \in A) \wedge \forall n, \exists \sigma_n, (s_n, s_{n+1}) \in f_{\sigma_n} \\
\equiv\ & (s_0 \in A) \wedge \exists \sigma, \forall n, (s_n, s_{n+1}) \in f_{\sigma_n} \\
\equiv\ & \exists \sigma, s \in \theta_\sigma(A, f).
\end{aligned}
$$

It is also clear that using the observation function, all sequences of a trace-parametrized point-level dynamics are mapped on this trace.

Proposition 4.16. *If (X, f) is a RDS observed through the $\mathcal{V} : X \times X \mapsto \mathcal{X}$, and $A \subseteq X$, then*

$$\forall s \in \theta_w(A, f), \mathcal{V}(s) = w.$$

Proof. By construction of $\mathcal{V}$ and Def. 4.14.

4.3 Symbolic Observation

In §4.1, we started with the introduction of the notion of observation. In the rest of this monograph, we will consider coarse-grained observations of systems, where transitions of a system are denoted by symbols of a finite alphabet.

We want to have a symbolic description of our relational dynamical system in order to be able to follow its evolution as

– it visits particular regions of the space, a region being a part of its domain, of its range, or any combination domain-range;
– it chooses specific transitions between states.

The first view generalizes symbolic dynamics. The second one is more typical in the study of transition systems and automata, in which transitions and codomains are as important as domains.

If we cover the relation $f \subseteq X \times X$ by a family of sets of $X \times X$, we can associate symbols to these sets, and relate them to either subspaces or subsystems. The first choice implies to keep track of the subspaces that are visited along the evolution. The second choice implies to record the subsystems that are activated during the same evolution.

The two perspectives are implemented using the same technical notion of covering, applied to the space where the relation is defined.

Definition 4.17 (Covering). *Let (X, f) be a RDS. A covering α of f is a finite set of subsets A_i of $X \times X$. To each set is associated a symbol out of an alphabet Σ, which allows to write $\alpha = \{A_i | i \in \Sigma\}$. The sets A_i must cover f, viz. $f \subseteq \cup_i A_i$.*

Remark 4.18. – For them to be useful, the alphabet and the covering should contain at least two elements.
– In Def. 4.17, there is no assumption of disjointness between the sets of a covering: they could overlap each other. Thus, for each $(x, y) \in f$, two situations could exist: there is a unique symbol $i \in \Sigma$ such that $(x, y) \in A_i$; or, there is a unique set $I \in \mathbb{P}(\Sigma)$ such that $\forall j \in I, (x, y) \in A_j$. In the second case, we associate a unique symbol to each element of $\mathbb{P}(\Sigma)$, which permits us to treat symbols and sets of symbols in a same way. Since Σ is finite, its power set is also finite. Working at the symbolic set-level is equivalent to transforming the covering into a partition (covering where sets are mutually disjoint). In the following, we will always assume that the correspondence between state pairs and symbols is functional.
– Note that the subsets A_i used to define a covering are nothing but relations defined on X.

In the following, we will use some extra notations for particular coverings, using our standard notations.

Definition 4.19 (Inverse and restricted coverings). *If α is a covering associated with the alphabet Σ, the inverse covering, the domain-restricted covering and the range-restricted covering are respectively given by*

$$
\begin{aligned}
\alpha^{-1} &= \{A_i^{-1} | A_i \in \alpha\}, \\
(B \to \alpha) &= \{(B \to A_i) | A_i \in \alpha\}, \\
(\alpha \leftarrow B) &= \{(A_i \leftarrow B) | A_i \in \alpha\}.
\end{aligned}
$$

These three coverings are also associated with Σ.

Finally, using the covering defined on a system, we can parametrize our set-transformers explicitly with symbols, which corresponds to the part of space visited, or to the activation of a specific transition. Let (X, f) be a RDS, observed on covering $\alpha = \{A_j | j \in \Sigma\}$. Symbol $j \in \Sigma$ represents a possible transition. The set-transformer f can be parametrized by this transition:

$$f_j(A) = (f \cap A_j)(A)$$

where A_j is interpreted as a relation. If $s \subseteq \Sigma$ is a set of possible transitions, we have:

$$f_s(A) = (f \cap (\cup_{j \in s} A_j))(A).$$

Trace-parametrized set-transformers and point-level dynamics are easily adapted to this formalism, following §4.2.

4.4 Abstraction of Systems

The observation of systems presented in §4.1 can range from full precision to total absence of information (e.g. constant observation function). Abstract, viz. simplified, observations of the evolutions of a given dynamical system can be preferred to their concrete or actual description, when

– useless details of the dynamics can be omitted;
– a fully precise computations is not possible, due to lack of appropriate computational or analytical tools;
– a simplified view of the system and its evolution is needed.

Intuitively, a dynamical system (Y, g) abstracts another system (X, f) when the trajectories of g represent simplified trajectories of f. Of course, the related observations can also be abstracted.

Definition 4.20 (Abstraction function). *An abstraction function between two spaces X and Y is a total surjective function $\mathcal{Z} : X \mapsto Y$.*

Definition 4.21 (Abstraction homomorphism). *The abstraction function $\mathcal{Z} : X \mapsto Y$ is an abstraction homomorphism between the RDS (X, f) and (Y, g) iff $\mathcal{Z}(\theta(X, f)) = \theta(Y, g)$. We denote this homomorphism by $f \Subset g$.*

Remark 4.22. – The condition $\mathcal{Z}(\theta(X, f)) = \theta(Y, g)$ entails that the following diagram commutes:

$$
\begin{array}{ccc}
X & \xrightarrow{\ f\ } & X \\
{\scriptstyle \mathcal{Z}}\big\downarrow & & \big\downarrow{\scriptstyle \mathcal{Z}} \\
Y & \xrightarrow[\ g\ ]{} & Y.
\end{array}
$$

– If (X, f) and (Y, g) are respectively observed through the observation functions $\mathcal{V} : X \times X \mapsto \mathcal{X}$ and $\mathcal{W} : Y \times Y \mapsto \mathcal{Y}$, the following diagram must also commute:

$$
\begin{array}{ccc}
X \times X & \xrightarrow{\;\;\mathcal{V}\;\;} & \mathcal{X} \\
\mathcal{Z} \downarrow & & \downarrow \mathcal{Z} \\
Y \times Y & \xrightarrow[\;\;\mathcal{W}\;\;]{} & \mathcal{Y}.
\end{array}
$$

– Both actual and observation diagrams are summarized by the following one:

$$
\begin{array}{ccc}
X & \xrightarrow{\;\;f\;\;} & X \\
\mathcal{Z} \downarrow \quad \mathcal{V} & \downarrow \mathcal{Z} & \downarrow \mathcal{Z} \\
Y & \xrightarrow[\;\;g\;\;]{\mathcal{W}} & Y.
\end{array}
$$

4.5 Qualitative Abstract Verification

An abstraction homomorphism $\mathcal{Z}$ between two systems f and g such that $f \Subset g$ permits to verify some dynamical properties. The aim of this section is to show that abstraction can help in verifying important families of properties such as invariance and reachability, i.e. finite-time attraction. We just sketch the main ideas informally; more details on invariance and attraction will be given in Chap. 5.

We work with the set of trajectories of a RDS (X, f) starting from A, $\theta(A, f)$. For example, here are two basic definitions.

Definition 4.23 (Invariance). *Given a set $P \in X$ and a set of state sequences S, an invariance property of f is a predicate*

$$
\begin{aligned}
\mathbb{A} &: \quad \mathbb{P}(X) \times \mathbb{P}(X^{\mathbb{N}}) \mapsto \mathbb{B} \\
s.t. &\quad \mathbb{A}(P, S) \equiv \forall s \in S, \forall n, s_n \in P
\end{aligned}
$$

where $\mathbb{B}$ denotes the set of Boolean *values $\{\mathbf{t}, \mathbf{f}\}$.*

Definition 4.24 (Reachability). *A reachability property of f is a predicate*

$$
\begin{aligned}
\mathbb{E} &: \quad \mathbb{P}(X) \times \mathbb{P}(X^{\mathbb{N}}) \mapsto \mathbb{B} \\
s.t. &\quad \mathbb{E}(P, S) \equiv \forall s \in S, \exists n, s_n \in P
\end{aligned}
$$

The only difference between these two properties, where $\mathbb{A}$ stands for "always" and $\mathbb{E}$ for "eventually", is the logical quantifier: universal in the first case, existential in the second one [15, 190]. Based on this, the two following equivalences hold [288].

Proposition 4.25 (Homomorphic verification). *Let $\mathcal{Z} : X \mapsto Y$ be an abstraction homomorphism between two RDS, (X, f) and (Y, g). If $P \in \mathbb{P}(X)$, $Q \in \mathbb{P}(Y)$, and $P = \mathcal{Z}^{-1}(Q)$, then*

$$\mathbb{A}(P, \theta(X, f)) \equiv \mathbb{A}(Q, \theta(Y, g))$$
$$\mathbb{E}(P, \theta(X, f)) \equiv \mathbb{E}(Q, \theta(Y, g)).$$

Proof. We prove the invariance property.

$\Rightarrow$

$$\text{Let us choose } t \in \theta(Y, g)$$
$$\because \quad \mathcal{Z}(\theta(X, f)) = \theta(Y, g)$$
$$\Rightarrow \quad \exists s \in \theta(X, f), \mathcal{Z}(s) = t$$
$$\because \quad \mathbb{A}(P, \theta(X, f))$$
$$\Rightarrow \quad \forall n, s_n \in P$$
$$\Rightarrow \quad \forall n, \mathcal{Z}(s_n) \in \mathcal{Z}(P)$$
$$\because \quad P = \mathcal{Z}^{-1}(Q) \Rightarrow Q = \mathcal{Z}(P)$$
$$\Rightarrow \quad \forall n, t_n \in Q.$$

$\Leftarrow$

$$\text{Let us choose } s \in \theta(X, f)$$
$$\because \quad \mathcal{Z}(\theta(X, f)) = \theta(Y, g)$$
$$\Rightarrow \quad \exists t \in \theta(Y, g), \mathcal{Z}(s) = t$$
$$\because \quad \mathbb{A}(Q, \theta(Y, g))$$
$$\Rightarrow \quad \forall n, t_n \in Q$$
$$\because \quad \text{monotonicity of } \mathcal{Z}^{-1}$$
$$\Rightarrow \quad \forall n, \mathcal{Z}^{-1}(t_n) \subseteq \mathcal{Z}^{-1}(Q)$$
$$\because \quad P = \mathcal{Z}^{-1}(Q) \text{ and } t_n = \mathcal{Z}(s_n)$$
$$\Rightarrow \quad \forall n, \mathcal{Z}^{-1}(\mathcal{Z}(s_n)) \subseteq P$$
$$\because \quad \text{Prop. 2.49}$$
$$\Rightarrow \quad \forall n, s_n \in P.$$

Proving the reachability property follows the same development, replacing "$\mathbb{A}$" by "$\mathbb{E}$", and "$\forall n$" by "$\exists n$".

We have just shown that reachability, i.e. existence, and invariance, i.e. universality, can be preserved by homomorphism under some rather weak assumptions. Ideally, we would like to extend this proof scheme to basically all properties on trajectories of systems without sacrificing too much of the simplicity of our assumptions. Results in this directions have been published in the context of simulation of transition systems [64, 206]. In particular, stated in our framework, it has been shown that if R and its inverse are abstraction homomorphisms between two RDS, and R preserves a set of primitive formulas in both directions, then R preserves the Hennessy-Milner logic based on this set in both directions [64].

4.6 Observation as Abstraction

The observation function defined in §4.1 induces an interesting abstraction homomorphism. Let us thus transform a concrete dynamical system (X, f) observed through $\mathcal{V} : X \times X \mapsto \mathcal{X}$ into an abstract homomorphic dynamical system (Y, g).

Given an observation function, from every state transition of f, we obtain an observed transition. Thus, the point-level nondeterministic dynamics can be transformed into an abstract dynamics.

By definition, $\theta(A, f)$ contains sequences of the form

$$s_0 \longrightarrow s_1 \longrightarrow s_2 \longrightarrow s_3 \longrightarrow \cdots$$

where each transition belongs to f, and can thus be translated using $\mathcal{V}$. We obtain a set of traces, that we regard as new state transition sequences.

If we introduce the three sets

$$
\begin{aligned}
Y &= \mathcal{X} \\
S &= \mathcal{V}(\theta(A, f)) \\
A_\mathcal{V} &= \{t_0 \in Y \mid \exists t \in S\}
\end{aligned}
$$

we get the abstract dynamics of f:

$$\theta(A_\mathcal{V}, g) = S.$$

By construction of g, $\mathcal{V}$ realizes an abstraction homomorphism between f and g.

4.7 Discussion

In this section, we present an important technique related to symbolic observations of dynamical systems: symbolic dynamics, and discuss the consequences and limitations of qualitative abstract verification.

4.7.1 Observation and Abstraction: Related Work

Two complementary notions have been presented in this chapter, namely observation and abstraction. Both are classical in many fields, including dynamical systems and program theory.

Our definition of observation function is very general: it permits both precise and coarse-grained observations of systems, as well as labeling states and transitions. Moreover, as we briefly mentioned in §4.6, it can be considered as an abstraction homomorphism. The fact that we map state transitions to labels generalizes the classical state labeling used in symbolic dynamics: we can analyze relational nondeterministic systems or unions of systems in a same unified setting (see also §4.7.2 for more details on symbolic dynamics).

Observation is presented in [19] for transition systems, and its particularization to domains of systems is well known in symbolic dynamics [139]. It is also similar to labeling in processes algebras [151, 222]. In [127], the observation spaces of hybrid systems are related to quantum automata, and the authors suggest to study hybrid systems in the light of quantum mechanics, as observation is a central issue in this theory.

Abstraction homomorphisms are commonly used in representation theory and abstract interpretation [163], automata theory [117], hybrid systems [244, 288], symbolic dynamical systems [143, 180], and semi-conjugated dynamical systems [9].

Finally, first elements of a general theory of simulation in the context of dynamical systems have been published in [272, 220], and interesting results on the impact of simulation and precision on the observed dynamics of continuous cellular automata have been described in [103, 106].

4.7.2 Symbolic Dynamics vs Astract Observation

Symbolic dynamics is a method for studying dynamical systems, based on a strong abstraction homomorphism between concrete systems and a class of abstract symbolic dynamical systems. Let us first recall how it works, and then compare it with the notions introduced before.

Classical Symbolic Dynamics. Given a system (X, f), its domain is partitioned into a finite number of regions $\{D_i \mid i \in \Sigma\}$. Each step of the evolution is associated with a symbol corresponding to the region visited at this step. Thus, each trajectory s produces a corresponding symbolic trace t: $s_n = f^n(s_0) \in D_i \Rightarrow t_n = i$. The technique is used with functions, where the history of a state s_0, i.e. the trajectory s, uniquely depends on this state. Of course, this is no longer true when dealing with relations, and the whole trajectory is needed to compute the corresponding symbolic trace.

To get interesting results on the function by abstract analysis, a *topological conjugacy* is required: a continuous bijection $\mathcal{Z}$ between states and symbolic sequences such that its inverse is also continuous (*homeomorphism*); an abstraction homomorphism with the *shift dynamical system* ρ or a *subshift*:

$\forall x, \mathcal{Z}(f(x)) = \rho(\mathcal{Z}(x))$. The shift is $(\rho, \{0,1\}^{\mathbb{Z}})$, where $\forall x \in \{0,1\}^{\mathbb{Z}}, \rho(x)_i = x_{i+1}$, and a subshift $(\rho, B \subset \{0,1\}^{\mathbb{Z}})$ is a closed ρ-invariant subset B of $\{0,1\}^{\mathbb{Z}}$ [248, 34]. Then, working with the shift on a space of symbolic sequences is far simpler than working with the original system. As an example, the number of n-periodic sequences of ρ is 2^n, which is not always immediate regarding the underlying f.

Observation and Abstraction. This example permits to clearly separate observation from abstraction. Each state is mapped to a symbol through a coarse-grained observation function. States are then abstracted onto traces of symbols.

This observation is a particular case of our observation function which assigns labels to state transitions instead of single states. Symbolic dynamics does not involve transitions between states and codomains but only domains of functions. Non-functional systems are not easy to observe using classical symbolic dynamics, whereas it is natural in our framework. The following example illustrate this: we want to observe the nondeterministic evolution of a union of two different systems defined on the same domain. The notion of covering permits to model this.

Example 4.26. Let f and g be two functions defined on $[0,1]$ by $f(x) = \frac{x}{4}$ and $g(x) = \frac{x}{3}$. The union $f \cup g$ can be traced using the following covering: $\Sigma = \{0,1\}$, $A_0 = f$, $A_1 = g$, $\alpha = \{A_0, A_1\}$. Notice that choosing $\alpha = \{A'_0, A'_1\}$ with $A'_1 = Dom(f) \times Rg(f)$ and $A'_2 = Dom(g) \times Rg(g)$ would not be very helpful since many transitions would then belong to both sets.

Relating traces to states is not always possible when dealing with relations. In this case, symbolic sequences must be associated with trajectories that are not necessarily deterministic. Given a trajectory, the shift will verify the second condition (i.e. the abstraction homomorphism); the first one (i.e. the homeomorphism state-trace) does not hold anymore.

Related Notions. Other notions can be related to symbolic dynamics [88, 139, 202, 326]: Markov partitions, Bernoulli shifts (with probabilities assigned to transitions) [250], traces in parallelism semantics [89, 216, 309], etc.

4.7.3 Qualitative Abstract Verification

In §4.5, we have shown that some dynamical properties can be studied at an abstract level and their results are still valid at the concrete level.

What are the consequences? Studying the abstract system instead of the actual one still provides some *qualitative information* on the *observations* we can made of the system. However, the results do not always hold about the system itself: the assumptions of Def. 4.21 and Prop. 4.25 are rather strong, but weaker assumptions would not lead to the same conclusions. In the following, and particularly in Chap. 5, we will focus on invariance and attraction properties of dynamical systems, based on observable symbolic evolutions. We will have to keep in mind that the conclusions are not always transferable as such to the actual underlying systems, although the abstraction homomorphism permits to establish interesting qualitative relationships between state sequences (instead of states) and symbol sequences.

Is this a severe limitation? From the physical point of view, this limitation seems quite natural because our observational capabilities are themselves limited to our own perception of reality. Dealing with abstract systems is thus closer to what we can actually achieve.

5. Invariance, Attraction, Complexity

On the basis of the mathematical framework (Chap(s). 2 and 3), we introduced abstract observation means (Chap. 4) that preserve qualitative dynamical properties, e.g. invariance and reachability (see §4.5). This chapter focuses on these two properties, and relates them to dynamical complexity of systems.

Naively speaking, dynamical complexity can be approached by the following questions: How do states evolve? How do they behave? Where do they go? What do they do when they get there? Complexity is opposed to "poor" behaviors like fixpoints or cycles, even though 2^{100}-periodic cycles could be rather complicated to study!

Finite trajectories are generally simple and easily characterized, compared to infinite evolutions from which complexity seems to emerge. Invariants are sets of states that have infinite internal histories. Their structure represents trajectories between inner states; it organizes and, thus, strongly influences the resulting dynamics. Attraction establishes relations between initial and final or asymptotic states of infinite histories. Most of the time, attraction and invariance are combined as the attractors are themselves invariant sets, and one speaks about stabilization [109, 91, 326, 9, 123].

Devaney isolated three essential features of chaotic systems on their invariants: unpredictability (Sensitivity to Initial Conditions), which means that small perturbations of initial conditions induce dramatically divergent histories, space-indecomposability (Topological Transitivity), expressing how the evolutions of some states spread over the whole space, and regularity (Density of Periodic Points), guaranteeing the existence of a periodic point arbitrarily close to any other state of the invariant [88]. In [181], Knudsen proposed another definition of chaos based on the conjunction of the first two properties only.

Systems are rarely observed with full precision, and a coarse-grained observation is often preferred, which can also influence their analysis: infinite traces are related to invariant states, and trace languages reflect invariant structures [248, 88, 285]. A useful observation level must be such that sufficiently many traces are observable. On the other hand, if the precision is too low, interesting patterns of the dynamics remain hidden; thus, the observation level must reduce the number of states related to particular traces.

F. Geurts: Abstract Compositional Analysis of Iterated Relations, LNCS 1426, pp. 95-131, 1998.
© Springer-Verlag Berlin Heidelberg 1998

When these complementary conditions, namely fullness and atomicity, are both satisfied, all ingredients of complexity are present in the dynamics: the system is Knudsen chaotic on its invariant.

The chapter is organized as follows: in §§5.1, 5.2, 5.3, we respectively define invariants, analyze the impact of their structure on dynamical complexity, and establish criteria for their structural properties; various attraction properties are defined in §5.4, and a taxonomy is proposed; in §5.5, we introduce Lyapunov-like criteria to prove attraction properties; in §5.6, we combine invariance and attraction; finally, we close the chapter in §5.7 with a discussion. Related notions are presented in this final discussion; hence, only a few useful references are given in §§5.1–5.6.

5.1 Invariance

The usual definition of invariant is a set of points in which the system *must* stay when iterated forward indefinitely [9, 305, 326]. In temporal logic and program semantics, a less restrictive definition is also used: a set of points in which the system *can* stay when iterated forward (or backward) indefinitely [99, 93, 98, 210, 19].

In this section, both potential and necessary invariants are investigated. We refine the qualitative invariance property introduced in §4.5 to propose different definitions of necessary and potential invariants: forward and backward, global (viz. two-way) and strong (viz. indefinitely two-way) invariants. Extreme invariants are computed by successive iterations, as solutions of fixpoint equations [306, 70, 197, 93].

5.1.1 Forward and Backward Invariance

Let us start from the qualitative property presented in §4.5 (see Def. 4.23). Let S be a set of sequences defined on X, and J be a subset of X. Then, invariance of J is expressed as follows:

$$\mathbb{A}(J, S) \equiv \forall s \in S, \forall n, s_n \in J.$$

The property J is invariant on a set of histories S iff every state of every particular history is in J.

The set S of histories can be $\theta(X, f)$ for any RDS (X, f), or any set of histories starting from states of the invariant J itself. From this, we arrive at two notions of invariance:

− J is a *necessary invariant* iff

$$\forall x \in J, \forall s \in \theta(x, f), \mathbb{A}(J, s);$$

– J is a *potential invariant* iff

$$\forall x \in J, \exists s \in \theta(x, f), \mathbb{A}(J, s).$$

The following propositions give equivalent properties, that we will use as definitions for these differents types of invariants (see also Fig. 5.1):

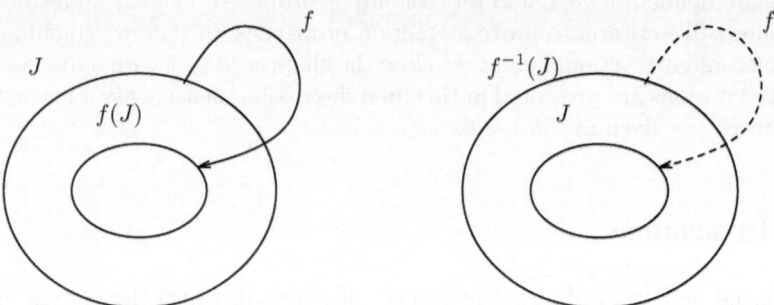

Fig. 5.1. Necessary invariant (left): from J, the images by f *must* (arrow) be in $f(J)$. Potential invariant (right): from $f^{-1}(J)$, the images by f *can* (dotted arrow) be in J.

– necessary invariant: $f(J) \subseteq J$, for $f(J)$ is the set of points reachable in one step from J;
– potential invariant: $J \subseteq f^{-1}(J)$, for $f^{-1}(J)$ is the set of points from which it is possible to reach J in one step.

Proposition 5.1 (Necessary invariance). *Let (X, f) be a RDS and $J \subseteq X$. Then,*

$$f(J) \subseteq J \Leftrightarrow \forall x \in J, \forall s \in \theta(x, f), \mathbb{A}(J, s).$$

Proof. $\Rightarrow$Let us consider $x \in J$ and $s \in \theta(x, f)$. By induction on n, we prove that $\forall n, s_n \in J$.
Basic case: $s_0 = x \in J$.
Inductive case: as $s \in \theta(x, f)$, by induction and $f(J) \subseteq J$, we have $s_{n+1} \in f(s_n) \subseteq f(J) \subseteq J$.
$\Leftarrow$Let y be in $f(J)$. There exists $x \in J, y \in f(x)$. As $\forall s \in \theta(x, f), \mathbb{A}(J, s)$, we have $y \in J$.

Remark 5.2. Either all histories of f are infinite or finite histories can be completed by the undefined state $\heartsuit$. Thus, as $y \in f(J)$, there is always a state $x \in J$ and a trajectory $s \in \theta(x, f)$ such that $s_0 = x$ and $s_1 = y$. If $f(y)$ is undefined and leads to a finite trajectory, this history is followed by an infinite sequence $\heartsuit^\omega$.

Proposition 5.3 (Potential invariance). *Let (X, f) be a RDS and $J \subseteq X$. Then,*

$$J \subseteq f^{-1}(J) \Leftrightarrow \forall x \in J, \exists s \in \theta(x, f), \mathbb{A}(J, s).$$

Proof. $\Rightarrow$Let us consider $x \in J$. We construct a sequence $s \in \theta(x, f), \mathbb{A}(J, s)$
by induction on n.

Basic case: $s_0 = x \in J$.

Inductive case: by induction, we have $s_n \in J$, and by $J \subseteq f^{-1}(J)$, $s_n \in$
$f^{-1}(J)$; thus, $\exists s_{n+1} \in J, (s_n, s_{n+1}) \in f$.

$\Leftarrow$Let x be in J. By hypothesis, $\exists s \in \theta(x, f), \mathbb{A}(J, s)$. Thus, $\exists s_1 \in J, (x, s_1) \in$
f and $x \in f^{-1}(J)$.

The following definition presents four cases mixing forward/backward evolutions with potential/necessary invariance.

Definition 5.4 (Invariants). *Let J be a subset of X, and (X, f) be a RDS.
Then, J is*

– *a necessary forward invariant iff*

$$f(J) \subseteq J;$$

– *a necessary backward invariant iff*

$$f^{-1}(J) \subseteq J;$$

– *a potential forward invariant iff*

$$J \subseteq f^{-1}(J);$$

– *a potential backward invariant iff*

$$J \subseteq f(J).$$

Remark 5.5. Some sets trivially verify these definitions: for instance, the
empty set verifies all of them, and the whole space X the first two ones.

If f is a monotonic set-transformer, which is always true for RDS (see
Prop. 2.35), it has been shown that the extreme solutions of these inequalities
can be obtained by successive iterations, as fixpoints of equivalent equations
[306, 70, 71, 93]. To find the least necessary invariant greater than A, we have
to compute an increasing sequence from A until it reaches a stationary point.
To find the greatest potential invariant smaller than B, a decreasing sequence
from B has to be computed until it reaches its stationary state. The set A
cannot be the bottom element of the lattice to produce a nontrivial invariant,
whereas B can be the top element, viz. X. The following proposition makes
use of transfinite iterations (see Def. 2.76); its proof is a simple application
of Theorems 2.44, 2.46, and 2.72.

Proposition 5.6 (Extreme invariants). *Let (X, f) be a dynamical system. The least necessary forward (backward) invariant greater than A, $J_f^+(A)$*

$(J_f^-(A))$, and the greatest potential forward (backward) invariant smaller than B, $J_+^f(B)$ $(J_-^f(B))$, are obtained as follows:

$$
\begin{array}{llllll}
LNFI: & A \subseteq f(A) & \Rightarrow & J_f^+(A) & = & \mathbb{S}f(A) & = & \cup_i f^i(A) \\
LNBI: & A \subseteq f^{-1}(A) & \Rightarrow & J_f^-(A) & = & \mathbb{S}f^{-1}(A) & = & \cup_i f^{-i}(A) \\
GPFI: & f^{-1}(B) \subseteq B & \Rightarrow & J_+^f(B) & = & \mathbb{G}f^{-1}(B) & = & \cap_i f^{-i}(B) \\
GPBI: & f(B) \subseteq B & \Rightarrow & J_-^f(B) & = & \mathbb{G}f(B) & = & \cap_i f^i(B).
\end{array}
$$

Notation 5.7. $-$ We use the upper-index (in $J^\pm$) for the necessary cases, and the lower-index (in $J_\pm$) for the potential cases, because "necessary" is stronger (or higher) than "potential".
$-$ In general, the complementary index (f) will be omitted when clear from the context.
$-$ The arguments A and B are left implicit if they stand for the whole space X.

Forward and backward invariants can be mixed, as well as necessary or potential ones. In the rest of this chapter, we will concentrate on potential invariants because they better encompass the relational nondeterminism. Necessary variants can be adapted easily.

The next proposition refines the nature of invariants in the context of RDS.

Proposition 5.8. Let (X, f) be a total nonempty RDS, then $J_-^f(X)$ and $J_+^f(X)$ are nonempty compact sets.

Proof. Let us prove that J_- is a nonempty compact set. The symmetric case of J_+ is based on the fact that f^{-1} is closed iff f is closed (Prop. 2.30).

$-$ J_- is compact: since X is compact and f is closed, the image of any closed subset $A \subseteq X$ is closed (Prop. 2.30), and $\forall n, f^n(X)$ is closed; as arbitrary intersections of closed sets are closed, $\cap_n f^n(X)$ is closed in X, whence compact.
$-$ J_- is nonempty:
  $-$ f is total $(Dom(f) = X)$ and closed; thus, the image of any nonempty compact set is nonempty, and $\forall n, f^n(X) \neq \emptyset$;
  $-$ consequently, $\forall n, \exists x_n \in f^n(X)$; since X is compact, $(x_n)_n$ accumulates in $x \in X$;
  $-$ this point x is such that $\forall n, x \in f^n(X)$: otherwise, $\exists k, x \notin f^k(X)$ and $\exists N_x, N_x \cap f^k(X) = \emptyset$ as $f^k(X)$ is closed; by monotonicity of f, $\forall j \geq k, f^j(X) = \emptyset$, which contradicts the accumulation in x. Hence, $x \in J_-$ which is not empty.

5.1.2 Global Invariance

The definitions introduced above are not symmetric in time. Here, we introduce global, i.e. two-way, invariants, which allows to consider systems and their inverses without any distinction.

Each state of a global invariant J has at least one infinite forward trajectory in J, and at least one infinite backward trajectory in J. Once the evolution direction is chosen, no change can happen.

Sometimes, to study specific systems, global invariants are used without really justifying this. Actually, it becomes useful when expanding and contracting behaviors are combined in the same time direction (e.g. see the analysis of the Smale horseshoe map in [326, §4.1, pp. 420–437] and §7.2).

The next definition generalizes this view: it suffices to consider all states that forward and backward invariants have in common.

Definition 5.9 (Global invariants). *Let (X, f) be a RDS. The least necessary global invariant greater than A, $J_f(A)$, and the greatest potential global invariant smaller than B, $J^f(B)$ are obtained as follows:*

$$
\begin{array}{llll}
LNGI: & A \subseteq f(A) \cap f^{-1}(A) & \Rightarrow & J_f(A) = J_f^-(A) \cap J_f^+(A) \\
GPGI: & f(B) \cup f^{-1}(B) \subseteq B & \Rightarrow & J^f(B) = J_-^f(B) \cap J_+^f(B).
\end{array}
$$

Remark 5.10. – These global invariants can be empty but they are always compact sets in case of RDS.
– If the symmetry is not useful in some particular case, the backward invariant can reduce to the whole space, which does not bring any modification in the intersection of the global invariant.

5.1.3 Strong Invariance

Let us now introduce a stronger type of invariant and compare it with the previous notion of global invariance.

Definition 5.11 (Strong invariants). *Let (X, f) be a RDS. The least necessary strong invariant greater than A, $\mathbb{J}_f(A)$, and the greatest potential strong invariant smaller than B, $\mathbb{J}^f(B)$, are respectively the least solution, greater than A, and the greatest solution, smaller than B, of of the fixpoint equations*

$$
\begin{array}{ll}
LNSI: & f(K) \cup f^{-1}(K) \subseteq K \\
GPSI: & K \subseteq f(K) \cap f^{-1}(K).
\end{array}
$$

provided $A \subseteq f(A) \cup f^{-1}(A)$ and $f(B) \cap f^{-1}(B) \subseteq B$.

Notation 5.12. *In the rest of this section, $\mathbb{J}$ will denote the greatest potential strong invariant smaller than X, i.e. $\mathbb{J}^f(X)$.*

This notion is stronger than Def. 5.9 because it requires that it is always possible to iterate forward and backward from any state of $\mathbb{J}$. This entails possibly "foolish" histories when the system combines forward and backward steps, whereas J only contains states that are potentially invariant under infinite forward or backward histories, without any change of time direction.

Proposition 5.13. *For any RDS (X, f), the greatest potential strong invariant is contained in the greatest potential global invariant:*

$$\mathbb{J} \subseteq J.$$

Proof. By definition, the strong invariant is such that $\mathbb{J} \subseteq f(\mathbb{J}) \cap f^{-1}(\mathbb{J})$. Thus, $\mathbb{J} \subseteq f(\mathbb{J})$ and $\mathbb{J} \subseteq J_-$ since J_- is the greatest such set. Symmetrically, $\mathbb{J} \subseteq J_+$, whence the result.

The following two lemmas guarantee that the strong invariant as defined by Def. 5.11 is a compact set: the underlying relation of its iterated set-transformer is closed and monotonic.

Lemma 5.14. *Let (X, f) be a RDS, then the set-transformer F defined as follows:*

$$\forall A \in X, F(A) = f(A) \cap f^{-1}(A)$$

is a closed relation.

Proof. Let $(x_n, y_n)_n$ be a sequence in F, converging to (x, y). Let us prove that this pair belongs to F.

For all n, we have $(x_n, y_n) \in F$ and $(x_n, y_n) \in f$. Since f is closed, $(x, y) \in f$. Symmetrically, $(x, y) \in f^{-1}$. Hence, $(x, y) \in F$.

Lemma 5.15. *Let (X, f) be a RDS, then the set-transformer F defined as follows:*

$$\forall A \in X, F(A) = f(A) \cap f^{-1}(A)$$

is monotonic.

Proof. Let us assume $A \subseteq B \subseteq X$, and pick $x \in F(A)$. Then,

$$
\begin{array}{ll}
x \in f^{-1}(A) & x \in f(A) \\
\Rightarrow \ \exists y \in A, x \in f^{-1}(y) & \Rightarrow \ \exists z \in A, x \in f(z) \\
\Rightarrow \ \exists y \in B, x \in f^{-1}(y) & \Rightarrow \ \exists z \in B, x \in f(z) \\
\Rightarrow \ x \in f^{-1}(B) & \Rightarrow \ x \in f(B)
\end{array}
$$

and $x \in F(B)$.

Proposition 5.16. *Let (X, f) be a RDS, then $\mathbb{J}$ is a compact set.*

Proof. Follows from Lemmas 5.14 and 5.15, as in Prop. 5.8.

Remark 5.17. The emptiness of invariants is not at all excluded.

Example 5.18. Let us consider $f = ([0, \frac{1}{2}] \rightarrow 2x) \cup ([\frac{1}{2}, 1] \rightarrow 2(1-x))$. If $\varepsilon < \frac{1}{10}$, then $F([\frac{1}{2}-\varepsilon, \frac{1}{2}+\varepsilon]) = [1-2\varepsilon, 1] \cap ([\frac{1}{4}-\frac{\varepsilon}{2}, \frac{1}{4}+\frac{\varepsilon}{2}] \cup [\frac{3}{4}-\frac{\varepsilon}{2}, \frac{3}{4}+\frac{\varepsilon}{2}]) = \emptyset$.

Strong invariants can be equivalent to global invariants under some restrictions on the underlying dynamical systems.

Proposition 5.19. *Let (X, f) be a RDS. If there exists a finite covering $\alpha = \{A_i\}$ such that each transition-parametrized set-transformer f_i is simple and injective, then the strong invariant $\mathbb{J}$ is equivalent to the global invariant $J = J_- \cap J_+$.*

Proof. Let us define $F(A) = f^{-1}(A) \cap f(A)$. We have to prove that $\forall n, F^n(X) = f^{-n}(X) \cap f^n(X)$. By induction on n, this is easy.

- The case $n = 0$ is straightforward.
- Assuming that the result holds for all $m \leq n$, we have to prove it for $n+1$.

$$
\begin{aligned}
&F^{n+1}(X) \\
=~ &F(F^n(X)) \\
=~ &f(F^n(X)) \cap f^{-1}(F^n(X)) \\
&\qquad \because \quad \text{by ind. hyp.} \\
=~ &f(f^n(X) \cap f^{-n}(X)) \cap f^{-1}(f^n(X) \cap f^{-n}(X)) \\
&\qquad \because \quad \text{finite union of simple injective relations} \\
=~ &\cup_{i,j} f_i(f^{-n}(X)) \cap f_i(f^n(X)) \\
&\cap f_j^{-1}(f^{-n}(X)) \cap f_j^{-1}(f^n(X)) \\
&\qquad \because \quad \begin{aligned} f_i(f^n(X)) \cap f_j^{-1}(f^{-n}(X)) \\ \subseteq f_i(f^{-n}(X)) \cap f_j^{-1}(f^n(X)) \end{aligned} \\
=~ &\cup_{i,j} f_i(f^n(X)) \cap f_j^{-1}(f^{-n}(X)) \\
=~ &\cup_i f_i(f^n(X)) \cap \cup_j f_j^{-1}(f^{-n}(X)) \\
=~ &f^{n+1}(X) \cap f^{-(n+1)}(X)
\end{aligned}
$$

Finally, Prop. 5.19 is easily adapted to the case of finite unions of injective relations.

Corollary 5.20. *Let (X, f) be a finite union of simple injective RDS. Then, the strong invariant $\mathbb{J}$ of f is equivalent to $J_- \cap J_+$.*

Proof. Direct application of Prop. 5.19, with a covering $\alpha = \{f_i\}$.

5.2 Structure of Invariants

The (dynamical) structure of an invariant is a description of all possible trajectories between its states. It determines the complexity of the dynamics of the system.

For instance, if an invariant contains a few isolated fixpoints, the dynamics on this invariant will be much simpler than the one resulting from a cycle of length 2^{100}, or from an uncountable Cantor set (Def. 2.28).

In general, it is not possible or not easy to study these trajectories with full precision. Thus, we propose to approximate these trajectories by observed histories, and to study the relations between invariant states and traces. Two complementary conditions establish the natural trade-off on the observation precision: fullness and atomicity. Sufficiently many traces must be observable, which limits the precision level, but the observation must be fine enough to provide useful information on the dynamics.

Below, we propose a trace-parametrization of invariants; we introduce the definitions of fullness and atomicity; we recall well-known definitions of chaos; finally, we show how they are related to fullness and atomicity, which prove thus fundamental for the characterization of complex dynamical phenomena.

Fullness and atomicity are classical in ergodic theory under the names of "exactness" and "weak generation" [248, 34, 196]. However, we follow here the technical presentation of [286, 290].

5.2.1 Trace-Parametrized Invariants

The observation we can make of a system strongly depends on the covering we define for it, i.e. the grain we choose for the observation. We have shown how set-transformers can be parametrized by symbols corresponding to observation coverings (see Def. 4.12). From this, the components of an invariant can be parametrized by traces of symbols, too.

Definition 5.21 (Trace-parametrized invariant). *The invariant* $J_{\sigma,\tau}$ *parametrized by the traces* σ *representing the past, and* τ *representing the future is defined by:* $\forall \sigma, \tau \in \Sigma^\omega$,

$$J_{\sigma,\tau} = J_{\sigma,} \cap J_{,\tau}$$

with partial invariants defined using one-way traces:
$$J_{\sigma,} = f_{\sigma,}(X)$$
$$J_{,\tau} = f_{\tau,}^{-1}(X).$$

Remark 5.22. – Notice the use of different notations $J_{\sigma,}$ and $J_{,\tau}$ for past and future traces (see also Not. 4.13).

– A trace-parametrized invariant $J_{\sigma,\tau}$ is a set of states having a possible past trace σ and a possible future trace τ. If these traces are infinite, the corresponding invariant contains states that remain in J as they evolve according to the traces.

– Equivalently, partial invariants can be defined by induction: $\forall i, j \in \Sigma$:

$$\begin{aligned} J_{\sigma i,} &= f_i(J_{\sigma,}) \\ J_{,j\tau} &= f_j^{-1}(J_{,\tau}) \end{aligned}$$

and

$$J_{\varepsilon,} = J_{,\varepsilon} = X.$$

Proposition 5.23. *The greatest potential global invariant is equal to the union of all possible bi-infinite trace-parametrized invariants:*

$$J = \cup_{\sigma,\tau \in \Sigma^\omega} J_{\sigma,\tau}.$$

Proof.

$$\begin{aligned} & J = \cup_{\sigma,\tau \in \Sigma^\omega} J_{\sigma,\tau} \\ \Leftarrow \quad & J_- \cap J_+ = (\cup_{\sigma \in \Sigma^\omega} J_{\sigma,}) \cap (\cup_{\tau \in \Sigma^\omega} J_{,\tau}) \\ & \quad \because \quad \text{symmetrically for the second terms} \\ \Leftarrow \quad & J_- = \cup_{\sigma \in \Sigma^\omega} J_{\sigma,} \\ & \quad \because \quad J_- = \cap_n f^n(X) \text{ and monotonicity of } f \\ \Leftarrow \quad & \forall n, f^n(X) = \cup_{\sigma \in \Sigma^n} J_{\sigma,} \\ \Leftarrow \quad & \text{Induction on } n : \\ (n = 0) \quad & f^0(X) = X = \cup_{\sigma \in \Sigma^0} J_{\sigma,} = J_\varepsilon, \\ (n > 0) \quad & f^{n+1}(X) = f(f^n(X)) = f(\cup_{\sigma \in \Sigma^n} J_{\sigma,}) \\ & = \cup_{i \in \Sigma} f_i(\cup_{\sigma \in \Sigma^n} J_{\sigma,}) = \cup_{\sigma \in \Sigma^{n+1}} J_{\sigma,} \end{aligned}$$

5.2.2 Fullness and Atomicity

In terms of trace-parametrized invariants, it is possible to characterize the structure of the global invariant J of a system f. The idea is to relate invariant states to observed infinite traces passing through these states.

The first property we introduce is a condition verifying the adequacy of the observation function used to follow trajectories inside the invariant. Sufficiently many traces must be observable. Here, we choose a totalistic version: we impose that all traces are observable. Thus, the trace language $L_t = \Sigma^{\mathbb{Z}}$, where Σ is the alphabet of the observation covering.

Definition 5.24 (Fullness). *The invariant J of f is full iff each bi-infinite trace is realizable:*

$$\forall \sigma, \tau \in \Sigma^\infty, \#J_{\sigma,\tau} \geq 1.$$

If f is associated with a single symbol, $\Sigma = 1$, there is a unique observable trace: $\cdots 111, 111 \cdots$ This is not very useful because all states of J are related to this unique trace, which motivates the introduction of the second property, guaranteeing a sufficiently precise observation. Again, our definition is strong: it imposes that at most one state corresponds to each trace.

Definition 5.25 (Atomicity). *The invariant J of f is atomic iff each bi-infinite trace determines at most one point in the state space:*

$$\forall \sigma, \tau \in \Sigma^\infty, \# J_{\sigma,\tau} \le 1.$$

Remark 5.26. If an "undefined" state $\heartsuit$ is added to the state space X, all traces are realizable. In this case, in order to keep the above definitions consistent and meaningful, fullness has to be adapted:

$$\forall \sigma, \tau \in \Sigma^\infty, (\# J_{\sigma,\tau} \ge 1) \wedge (J_{\sigma,\tau} \ne \{\heartsuit\});$$

and atomicity must be modified, too:

$$\forall \sigma, \tau \in \Sigma^\infty, (\exists x \in X, J_{\sigma,\tau} = \{x\}) \vee (J_{\sigma,\tau} = \{\heartsuit\}).$$

Example 5.27 (A fixpoint invariant). Consider the following function whose domain is restricted to $[0,1]$ and whose range is restricted to $[0,2]$,

$$f_1(x) = ([0,1] \to 2x \leftarrow [0,2]).$$

Let us compute successive iterations leading to the potential invariant J_+:

$$
\begin{aligned}
f^{-1}([0,2]) &= [0,1] \\
f^{-2}([0,2]) &= f^{-1}([0,1]) = [0,\tfrac{1}{2}] \\
f^{-3}([0,2]) &= f^{-1}([0,\tfrac{1}{2}]) = [0,\tfrac{1}{4}] \\
&\;\;\vdots \\
f^{-i}([0,2]) &= [0,\tfrac{1}{2^{i-1}}] \\
J_+ &= \cap_i f^{-i}([0,2]) = \cap_i [0,\tfrac{1}{2^{i-1}}] = \{0\}.
\end{aligned}
$$

In the same way, we compute J_-:

$$
\begin{aligned}
f([0,1]) &= [0,2] \\
f^2([0,1]) &= f([0,2]) = [0,2] \\
&\;\;\vdots \\
f^i([0,1]) &= [0,2] \\
J_- &= \cap_i f^i([0,1]) = [0,2].
\end{aligned}
$$

The system considered here is basic. We associate a covering containing a unique set, i.e. $\alpha = \{f_1\}$, and the alphabet is $\Sigma = \{1\}$. There is only one bi-infinite trace, viz. $\cdots 111, 111 \cdots$, and one state in the invariant $J = \{0\}$.

Let us add another expanding transition f_2, also defined on $[0, 1]$:

$$f_2(x) \quad = \quad ([0, 1] \to 3x \leftarrow [0, 3])$$
$$f \quad = \quad f_1 \cup f_2.$$

The covering is extended to $\alpha' = \{f_1, f_2\}$. Symbols 1 and 2 are respectively associated with the activation of transitions f_1 and f_2. The alphabet is $\Sigma' = \{1, 2\}$. The invariant is still the singleton $\{0\}$, but the set of traces is uncountable ($\{1, 2\}^{\mathbb{Z}}$): there is one constant orbit but many different traces. The invariant is trivially full: every bi-infinite trace determines the same unique fixpoint $x = 0$.

Because of the time-symmetry present in our definitions, the same properties hold for the inverse system

$$f^{-1}(x) = ([0, 2] \to \frac{x}{2} \leftarrow [0, 1]) \cup ([0, 3] \to \frac{x}{3} \leftarrow [0, 1]).$$

This example shows that working at the level of trace can entail strange conclusions: in this case, the trace behavior seems very rich, whereas there is a unique attracting fixpoint. However, they correspond to the *observed* reality.

5.2.3 Chaos

We propose to study the structure of invariants with fullness and atomicity. Based on symbolic observations of evolutions, we will show that they entail important topological properties related to a widespread notion of complexity, namely *chaos*, under some conditions. Thus, fullness and atomicity permit to characterize some complex behaviors. First, let us recall Devaney's definition of chaos [88].

Definition 5.28 (Devaney chaos). *Let (X, f) be a simple RDS defined on a metric space (X, d). If there is a necessary forward invariant set J such that $f(J) \subseteq J$, f is Devaney chaotic on J iff*

– *it is sensitive to initial conditions,*

$$\exists \delta > 0, \forall x \in J, \varepsilon > 0, \exists y, n, \quad (d(x, y) < \varepsilon) \\ \wedge \quad (d(f^n(x), f^n(y)) > \delta); \tag{SIC}$$

– *it is topologically transitive,*

$$\forall U \text{ open}, V \subseteq J, \exists k \in \mathbb{N}, f^k(U) \cap V \neq \emptyset; \tag{TT}$$

– *periodic points are dense,*

$$\forall x \in J, \varepsilon > 0, \exists y, \quad (d(x, y) < \varepsilon) \\ \wedge \quad (\exists p > 0, f^p(y) = y). \tag{DPP}$$

Remark 5.29. – These three properties are important as such because they constitute basic blocks that can be found in many global characterizations of complexity [248, 34, 196, 285]. For instance, ergodic theory shows the same kind of notions with mixing, etc.

– TT is characterized by the existence of a dense orbit, i.e.

$$\exists x \in X, \forall y \in X, \forall \varepsilon > 0, \exists n, d(f^n(x), y) < \varepsilon.$$

– SIC is redundant, as proved in [26], i.e.

$$TT \wedge DPP \Rightarrow SIC$$

and no other pair of properties implies the third one in general [24].

– TT implies both DPP and SIC when functions are defined on intervals [310].

In [181], it is argued that DPP is not essential to obtain or characterize complex behaviors, whence the author proposes the following weaker definition.

Definition 5.30 (Knudsen chaos). *Let (X, f) be a simple RDS defined on a metric space (X, d). If there is a necessary forward invariant set J such that $f(J) \subseteq J$, f is Knudsen chaotic on J iff both SIC and TT hold.*

Corollary 5.31. *Devaney chaos implies Knudsen chaos.*

Usually, the way to prove SIC, TT and DPP is by finding a topological conjugacy with a symbolic dynamical system like the full shift or a subshift (see §4.7.2), since it is easier to prove properties for symbolic dynamical systems than for their original counterpart, and the conjugacy preserves all topological properties from symbolic, viz. abstract, to concrete systems. For instance, the following proposition is proved using symbolic dynamics [326].

Proposition 5.32. *If there exists a topological conjugacy between a system on its invariant and a full shift or a subshift, both abstract, viz. symbolic, and concrete systems have*

1. *a dense countable infinity of periodic points, consisting of points of all periods ($\Rightarrow$ DPP);*
2. *an uncountable infinity of nonperiodic points;*
3. *a dense orbit ($\Rightarrow$ TT, Rem. 5.29).*

Remark 5.33. We prove these properties for a general class of subshifts in Chap. 8 (see Theorem 8.14).

Corollary 5.34. *Using the assumptions of Prop. 5.32, both systems are Devaney chaotic on their respective invariants.*

5.2.4 Fullness Implies Trace Chaos

Fullness entails chaos w.r.t. traces. The following proposition rephrases the properties of Prop. 5.32 in terms of traces instead of points.

Proposition 5.35 (Trace chaos). *If the invariant of a RDS is full, and the covering is associated with an alphabet Σ containing at least two symbols, then the set of bi-infinite traces $(\sigma, \tau) \in \Sigma^{\mathbb{Z}}$ contains*

1. *a dense countable infinity of periodic traces consisting of traces of all periods;*
2. *an uncountable infinity of nonperiodic traces;*
3. *a dense trace (contains all arbitrarily large finite traces), $\Rightarrow$ TT.*

Proof. Once we know that all traces are realizable, by fullness of the invariant, these properties are trivially verified for the language of all traces $L_t = \Sigma^{\mathbb{Z}}$.

These three properties are related to traces, viz. observed trajectories, instead of points. They are thus weaker than the three similar properties presented in Prop. 5.32.

Proposition 5.36 (Fairness). *Fullness of the invariant implies fairness of the system.*

Proof. Fairness [110] means that no component of a structured systems can be forgotten forever. Fullness says that all traces are realizable, including the traces where no component is omitted forever. Thus, fair traces are contained in the structure of any full invariant.

Remark 5.37. The converse is not true in general. For this, the definition of fullness should be generalized, by using higher-order languages of feasible traces [72, 88, 116, 170, 322, 326, 333, 334]: subshifts, context-free, context-sensitive, or general languages; here, we simply take the regular language containing all bi-infinite traces.

5.2.5 Fullness and Atomicity Imply Knudsen Chaos

When both fullness and atomicity properties are verified, they entail two properties of chaotic systems: topological transitivity and sensitivity to initial conditions.

Topological transitivity is obtained as follows.

Proposition 5.38 (Topological transitivity). *If the covering $\alpha = \{A_i \mid i \in \Sigma\}$ associated with a dynamical system (X, f) is such that each f_i is injective, then fullness implies that any part of the invariant J of f can be reached from any other part in finitely many iterations:*

$$\forall \sigma_1, \sigma_2, \tau_1, \tau_2 \in \Sigma^*, f_{\tau_1 \sigma_2}(J_{\sigma_1, \tau_1}) \cap J_{\sigma_2, \tau_2} \neq \emptyset$$

Fullness and atomicity entail topological transitivity.

Proof. Fullness implies that $\forall \sigma_1, \sigma_2, \tau_1, \tau_2 \in \Sigma^*, \exists \sigma, \tau \in \Sigma^\omega, J_{\sigma\sigma_1, \tau_1\sigma_2\tau_2\tau} \neq \emptyset$. Moreover, as $\forall i, f_i$ is injective, we have $f_{\tau_1\sigma_2}(J_{\sigma\sigma_1, \tau_1\sigma_2\tau_2\tau}) = f_{\tau_1\sigma_2}(J_{\sigma\sigma_1,}) \cap f_{\tau_1\sigma_2}(J_{,\tau_1\sigma_2\tau_2\tau})$. Thus,

$$J_{\sigma\sigma_1, \tau_1\sigma_2\tau_2\tau} \subseteq J_{\sigma_1, \tau_1} \subseteq J$$

$$\because \quad \text{monotonicity and injectivity}$$

$$\Rightarrow \quad J_{\sigma\sigma_1\tau_1\sigma_2, \tau_2\tau} \subseteq f_{\tau_1\sigma_2}(J_{\sigma_1, \tau_1})$$

$$\because \quad J_{\sigma\sigma_1\tau_1\sigma_2, \tau_2\tau} \subseteq J_{\sigma_2, \tau_2}$$

$$\Rightarrow \quad (f_{\tau_1\sigma_2}(J_{\sigma_1, \tau_1})) \cap J_{\sigma_2, \tau_2} \neq \emptyset.$$

Atomicity adds that these parts of J can be as small as desired.

The relationship between sensitive dependence on initial conditions and fullness and atomicity is given by the following proposition.

Proposition 5.39 (Sensitivity to initial conditions). *Fullness   and atomicity entail sensitive dependence on initial conditions.*

Proof. Take two distinct states x and y in the invariant J. Atomicity implies a kind of contraction: there is at most one point in every bi-infinite invariant. Fullness guarantees that these sub-invariants are never empty. Since $x \neq y$, there exist $\sigma, \sigma', \tau, \tau'$ such that $x = J_{\sigma, \tau}$ and $y = J_{\sigma', \tau'}$. These bi-infinite traces are different and the first place where they differ gives the n we need to make the iterations diverge.

Remark 5.40. The symmetry of the systems we work with allows us to consider a sensitive dependence on final conditions, too.

Corollary 5.41 (Knudsen chaos). *If the covering $\alpha = \{A_i \mid i \in \Sigma\}$ associated with a dynamical system (X, f) is such that each f_i is injective, then fullness and atomicity of the global invariant J^f imply Knudsen chaos.*

5.2.6 Devaney vs Trace vs Knudsen Chaos

Let us summarize the essential features of the previous definitions and properties.

- The existence of a topological conjugacy with a symbolic shift entails TT and DPP, whence SIC, by Prop(s). 5.29 and 5.32. Thus, by Def. 5.28, the system is *Devaney chaotic* at the level of states. We call this *state chaos*.
- Fullness entails *trace chaos*: existence of a dense trace, and density of periodic traces (Prop. 5.35); these properties rephrase TT and DPP at the level of traces.
- When atomicity is added, TT and SIC are verified (Prop(s). 5.38 and 5.39). However, Devaney chaos is not present because DPP is not verified; there is only density of periodic traces. Nonetheless, as stated in Cor. 5.41, the system is *Knudsen chaotic* at the level of states, by Def. 5.30.

Using the result of Prop. 5.23, we can go one step further. Abstracting invariant states into traces requires a total surjective function (see Def. 4.20): $\mathcal{Z} : J \mapsto \Sigma^{\mathbb{Z}}$ such that $\mathcal{Z}(x) = \{(\sigma, \tau) \in \Sigma^{\mathbb{Z}} \mid x \in J_{\sigma,\tau}\}$.

On one hand, symbolic dynamics establishes more: a homeomorphic total bijective function from invariant states to traces. On the other hand, fullness induces a surjective relation, and atomicity entails an injective relation.

The functional and total characters of the abstraction function are present in the topological conjugacy but not in the conjunction of both fullness and atomicity. However, Prop. 5.23 states that $\cup_{(\sigma,\tau) \in \Sigma^{\mathbb{Z}}} J_{\sigma,\tau} = J$, which entails surjectivity from traces to invariant sets, or a total function from states to traces. Consequently, it is possible to prove that the countable infinity of periodic points holds at both abstract and concrete levels.

Proposition 5.42 (Countable infinity of periodic points). *If the potential invariant J of a simple injective RDS (X, f) is full and atomic, it has a countable infinity of periodic points.*

Proof. Fullness implies the existence of a countable infinity of periodic traces. Each states determines exactly one trace in the future and in the past, since the system is simple and injective, that is, deterministic in both time directions. Adding atomicity entails that exactly one state corresponds to each trace. Moreover, each periodic trace is associated with a periodic state. Thus, the number of periodic states equals the number of periodic traces.

The important difference between (Devaney) state chaos and trace chaos is the following. Fullness and atomicity are not strong enough to preserve the density of this countable infinity of periodic traces at the level of states. This role is played by the homeomorphic (viz. continuity in both directions) aspect of the function involved in the topological conjugacy. Abstraction alone is not sufficient to preserve such topological properties.

5.3 Fullness and Atomicity Criteria

The aim of this section is to develop criteria allowing to prove fullness and atomicity of particular systems, instead of having to use the definitions of these properties. We first generalize criteria given in [286, 290], and we show their application to three classical examples.

5.3.1 Criteria

The following proposition gives sufficient criteria for fullness. The main idea consists in the approximation of trace-parametrized invariants from below.

Proposition 5.43 (Fullness criteria). *The fullness of the invariant J of a system f can be proved by finding two sets of sets Φ and Ψ verifying the following criteria:*

1. $\forall i \in \Sigma, (\exists A \in \Phi, A \subseteq J_{i,}) \wedge (\exists B \in \Psi, B \subseteq J_{,i});$
2. $\forall i \in \Sigma, (\forall A \in \Phi, \exists A' \in \Phi, A' \subseteq f_i(A))$
   $\wedge (\forall B \in \Psi, \exists B' \in \Psi, B' \subseteq f_i^{-1}(B));$
3. $\forall A \in \Phi, B \in \Psi, A \cap B \neq \emptyset.$

The sets A, B in Φ, Ψ approximate the components J_σ, and $J_{,\tau}$ from below; if their intersection $A \cap B$ is never empty, then no $J_{\sigma,\tau} = J_\sigma, \cap J_{,\tau}$ is empty either.

Lemma 5.44. *If there exist two sets of sets Φ and Ψ such that criteria (1) and (2) of Prop. 5.43 are verified, then*

$$\forall n \geq 1, \forall \sigma, \tau \in \Sigma^n, \exists A \in \Phi, B \in \Psi, (A \subseteq J_\sigma,) \wedge (B \subseteq J_{,\tau}).$$

Proof. We prove this by induction on n:

$- n = 1$, this is given by (1);
$- n > 1$, suppose the property is valid up to n, let us prove it for $n + 1$:

$$\sigma \in \Sigma^n$$
$$\Rightarrow \quad i\sigma \in \Sigma^{n+1}$$
$$\wedge \exists A \in \Phi, A \subseteq J_\sigma,$$
$$\because \quad \text{monotonicity}$$
$$\Rightarrow \quad f_i(A) \subseteq f_i(J_\sigma,)$$
$$\Rightarrow \quad f_i(A) \subseteq J_{\sigma i,}$$
$$\because \quad (2)$$
$$\Rightarrow \quad \exists A' \in \Phi, A' \subset J_{\sigma i,}$$

and the same applies to $J_{,j\tau}$, symmetrically.

From this, we prove Prop. 5.43.

Proof. Taking (3) together with the previous lemma entails the result.

Now, we examine the case of atomicity. Here, we approximate the components of the invariant from above. As the number of states related to particular traces must be reduced to at most one when these traces indefinitely grow, a *Lyapunov-like decreasing function* is used: its value has to decrease at each improvement of the approximation of a trace-parametrized invariant, and its minimal value implies the atomicity condition.

Notation 5.45. *In the following, we use two obvious notations:*

$- \Phi \cap \Psi = \{A \cap B \mid A \in \Phi, B \in \Psi\};$

– $\cap_\Phi$ designates the smallest set of Φ such that the stated properties are verified. Here, we want the result of the intersection to be in Φ. If it is not, we take the smallest set of Φ containing this intersection.

Proposition 5.46 (Atomicity criteria). *Atomicity can be detected by finding two sets of sets Φ and Ψ, and a bounded function $\mathcal{H}$ from $\Phi \cap \Psi$ to an ordered set M with minimum 0, which verify the following:*

1. $\forall i \in \Sigma, (\exists A \in \Phi, J_{i,} \subseteq A) \wedge (\exists B \in \Psi, J_{,i} \subseteq B)$;
2. $\forall i \in \Sigma, (\forall A \in \Phi, f_i(A) \in \Phi) \wedge (\forall B \in \Psi, f_i^{-1}(B) \in \Psi)$;
3. $\forall A \in \Phi, B \in \Psi, \mathcal{H}(A \cap B) = 0 \Rightarrow \#(A \cap B) \leq 1$;
4. $\exists k, 0 < k < 1, \forall i, j \in \Sigma, \forall A \in \Phi, B \in \Psi,$
   $\mathcal{H}(f_i(A) \cap f_j^{-1}(B)) \leq k \cdot \mathcal{H}(A \cap B).$

Proof.

$$\because \quad (1)$$
$$\Rightarrow \quad \forall \sigma, \tau \in \Sigma^\infty, \forall n, \exists A \in \Phi, B \in \Psi, (J_{\sigma_n,} \subseteq A) \wedge (J_{,\tau_n} \subseteq B)$$
$$\because \quad (2), (4)$$
$$\Rightarrow \quad \mathcal{H}(f_{\sigma|_n,}(A) \cap f_{,\tau|_n}^{-1}(B)) \leq k^{n-1} \mathcal{H}(A \cap B) \leq k^{n-1} C$$
$$\because \quad A_n = \cap_\Phi \{A \mid A \in \Phi, J_{n,} \subseteq A\}$$
$$\because \quad \widetilde{A} = \cap_\Phi \{A \mid A \in \Phi, A \supseteq \cup A_n\}$$
$$\because \quad \text{idem for } \widetilde{B}$$
$$\Rightarrow \quad \lim_{n \to \infty} \mathcal{H}(f_{\sigma|_n,}(\widetilde{A}) \cap f_{,\tau|_n}^{-1}(\widetilde{B})) = 0$$
$$\because \quad (3)$$
$$\Rightarrow \quad \#(f_\sigma,(\widetilde{A}) \cap f_{,\tau}^{-1}(\widetilde{B})) \leq 1$$
$$\because \quad \forall n, J_{\sigma|_n,} \subseteq f_{\sigma|_{n-1},}(A) \subseteq f_{\sigma|_{n-1},}(\widetilde{A})$$
$$\Rightarrow \quad \#J_{\sigma,\tau} \leq 1.$$

Remark 5.47 (Atomicity criteria). – Instead of the last condition, a discrete version can also be used: $\forall i, j \in \Sigma, A \in \Phi, B \in \Psi,$

$$(\mathcal{H}(f_i(A) \cap f_j^{-1}(B)) = 0) \vee (\mathcal{H}(f_i(A) \cap f_j^{-1}(B)) < \mathcal{H}(A \cap B)).$$

– These conditions respectively correspond to Lyapunov stability functions [120, 119, 221] and to Floyd termination functions [109, 91]; they use the nonnegative reals and the naturals as set M.
– Using the notation $\widetilde{A} = \cap_\Phi \{B \mid B \in \Phi, B \supseteq A\}$, atomicity criteria become:
1. $\forall i \in \Sigma, (\widetilde{J_{i,}} \in \Phi) \wedge (\widetilde{J_{,i}} \in \Psi)$;
2. $\forall i \in \Sigma, (\forall A \in \Phi, \widetilde{f_i(A)} \in \Phi) \wedge (\forall B \in \Psi, \widetilde{f_i^{-1}(B)} \in \Psi)$;
3. $\forall A \in \Phi, B \in \Psi, \mathcal{H}(A \cap B) = 0 \Rightarrow \#(A \cap B) \leq 1$;
4. $\exists k, 0 < k < 1, \forall i, j \in \Sigma, \forall A \in \Phi, B \in \Psi,$
   $\mathcal{H}(\widetilde{f_i(A)} \cap f_j^{-1}(B)) \leq k \cdot \mathcal{H}(A \cap B).$

Actually, in this case, (2) is weaker than in Prop. 5.46 but (4) is stronger.

These sufficient criteria for fullness and atomicity can be generalized to include more cases, which inevitably leads to more complicated expressions. We refer the interested reader to [116] for a detailed study of some generalizations of the concepts presented above.

5.3.2 Case Studies: Dyadic Map, Cantor Relation, Logistic Map

Here, we rederive well-known results on the dyadic map, the Cantor relation and the logistic map, to illustrate the use of the criteria proposed for fullness and atomicity.

Example 5.48 (Dyadic map). Let the dyadic map be defined as follows (see Fig 5.2):

$$f = f_1 \cup f_2$$
$$f_1(x) = ([0, \frac{1}{2}] \to 2x \leftarrow [0, 1])$$
$$f_2(x) = ([\frac{1}{2}, 1] \to 2x - 1 \leftarrow [0, 1]).$$

Some easy computations lead to the global invariant:

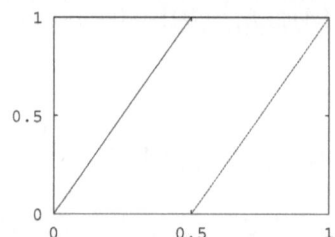

Fig. 5.2. Dyadic map $2x$ mod 1

$$f^{-1}([0, 1]) = [0, \frac{1}{2}] \cup [\frac{1}{2}, 1] = [0, 1]$$
$$J_+ = [0, 1]$$

and

$$f([0, 1]) = [0, 1] \cup [0, 1] = [0, 1]$$
$$J_- = [0, 1].$$

Fullness. To apply Prop. 5.43, we choose the approximating sets

$$\begin{aligned} \varPhi &= \{[0,1]\} \\ \varPsi &= \{[p,q] \mid 0 \le p \le q \le 1\}. \end{aligned}$$

Three conditions mustt be checked.

1. Immediate: for $i = 1, 2$,

$$\begin{aligned} J_i, &= f_i([0,1]) = [0,1] \in \varPhi \\ J_{,1} &= f_1^{-1}([0,1]) = [0, \tfrac{1}{2}] \in \varPsi \\ J_{,2} &= f_2^{-1}([0,1]) = [\tfrac{1}{2}, 1] \in \varPsi. \end{aligned}$$

2. Immediate w.r.t. $\varPhi$. Regarding $\varPsi$, we have

$$\begin{aligned} f_1^{-1}([p,q]) &= [\tfrac{p}{2}, \tfrac{q}{2}] \in \varPsi \\ f_2^{-1}([p,q]) &= [\tfrac{p+1}{2}, \tfrac{q+1}{2}] \in \varPsi. \end{aligned}$$

3. This condition holds since $[0,1] \cap [p,q] \ne \emptyset$, given $0 \le p \le q \le 1$. Thus, each component $J_{\sigma,\tau}$ contains at least one state, and J is full. Moreover, since $J = [0,1]$, each point in $[0,1]$ is contained in some $J_{\sigma,\tau}$.

Note the exact expression of these components takes the following form, for $\sigma, \tau \in \Sigma^n$, and $0 \le m < 2^n$:

$$J_{\sigma,\tau} = [\frac{m}{2^n}, \frac{m+1}{2^n}].$$

The use of the approximating sets $[p,q]$ allows to prove fullness without having to compute these exact expressions explicitly, which might reveal unfeasible in practice.

Atomicity. We prove it using the sufficient criteria of Prop. 5.46, with $\varPhi$ and $\varPsi$ defined as above, and

$$\mathcal{H}([p,q]) = q - p.$$

1. Verified as above for fullness.
2. Similar to the previous one.
3. Holds since $q - p = 0$ entails $\#\{x \mid p \le x \le q\} = 1$.
4. Verified as follows, using $k = \tfrac{1}{2}$:

$$\begin{aligned} \mathcal{H}(f_1([0,1]) \cap f_1^{-1}([p,q])) &= \mathcal{H}([\tfrac{p}{2}, \tfrac{q}{2}]) \\ &= \frac{q}{2} - \frac{p}{2} \\ &= \frac{1}{2} \cdot \mathcal{H}([p,q]), \end{aligned}$$

and similarly for the other cases

$$\mathcal{H}(f_i([0,1]) \cap f_j^{-1}([p,q])).$$

Each $f_i(A) \cap f_j^{-1}(B)$ decreases the size of $A \cap B$ at least by $\frac{1}{2}$. Each limit component $J_{\sigma,\tau}$ (with $\sigma, \tau \in \Sigma^\infty$) contains at most one state: J is atomic. Since J is also full, each $J_{\sigma,\tau}$ contains exactly one state. Thus, by Prop(s). 5.38 and 5.39, the dyadic map is topologically transitive and sensitive to initial conditions.

Example 5.49 (A Cantor-set invariant). Let us first consider $f_1(x) = \frac{x}{3}$ on $[0,1]$. Its positive invariant J_+ is $[0,1]$, its negative invariant J_- is $\{0\}$, and its global invariant J is $\{0\}$. Since there is only one function, we only attach one symbol to it, 1, and the invariant is trivially full and atomic, because there is only one bi-infinite trace $\cdots 111, 111 \cdots$ and one trace-parametrized invariant which is equal to the single point $\{0\}$.

We now add $f_2(x) = \frac{2}{3} + \frac{x}{3}$ to the system and take the union $f_1 \cup f_2$ (see Fig. 5.3). Then the positive invariant is again $[0,1]$ but the negative invariant

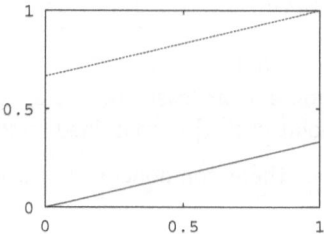

Fig. 5.3. Cantor relation

is the well-known Cantor middle-thirds Set (see Ex. 2.29). We can see this by computing the first few iterates leading to J_-, i.e. $f^i([0,1])$:

$$f([0,1]) = [0,\tfrac{1}{3}] \cup [\tfrac{2}{3},1]$$

$$f^2([0,1]) = f([0,\tfrac{1}{3}] \cup [\tfrac{2}{3},1])$$

$$= [0,\tfrac{1}{9}] \cup [\tfrac{2}{9},\tfrac{3}{9}] \cup [\tfrac{6}{9},\tfrac{7}{9}] \cup [\tfrac{8}{9},1]$$

$$\vdots$$

This invariant is more difficult to characterize, because limit points of this process are numerous, and it is not clear that each bi-infinite trace defines a unique state of the Cantor-set invariant. Using the criteria proposed above, and the covering $\alpha = \{f_1, f_2\}$ associated with $\Sigma = \{1,2\}$, let us show that this invariant is full and atomic. The analysis is very similar to the previous one (dyadic map, Ex. 5.48).

Fullness. To apply Prop. 5.43, we have to choose two sets of sets of $[0, 1]$ approximating iterations from below:

$$\begin{aligned} \Phi &= \{[p, q] \mid 0 \le p \le q \le 1\} \\ \Psi &= \{[0, 1]\}. \end{aligned}$$

Now, let us verify each condition.

1. Each $J_{i,}$ is in Φ:

$$\begin{aligned} J_1, &= f_1([0, 1]) = [0, \frac{1}{3}] \\ J_2, &= f_2([0, 1]) = [\frac{2}{3}, 1] \end{aligned}$$

and each $J_{,j}$ is obviously in Ψ:

$$J_{,j} = f_j^{-1}([0, 1]) = [0, 1].$$

2. For any $A = [p, q] \in \Phi$, each $f_i(A) \in \Phi$:

$$\begin{aligned} f_1([p, q]) &= [\frac{p}{3}, \frac{q}{3}] \\ f_2([p, q]) &= [\frac{p+2}{3}, \frac{q+2}{3}] \end{aligned}$$

and for any $B \in \Psi$ (there is only one such B), each $f_j^{-1}(B) \in \Psi$:

$$f_j^{-1}(B) = f_j^{-1}([0, 1]) = [0, 1].$$

3. Finally, any intersection of a set $A \in \Phi$ with a set $B \in \Psi$ is nonempty:

$$\forall 0 \le p \le q \le 1, [p, q] \cap [0, 1] = [p, q] \ne \emptyset.$$

From this, we conclude the the global invariant of this system is full.

Atomicity. The approximating sets of fullness can also help here. We add a decreasing function

$$\mathcal{H}([p, q]) = q - p.$$

Let us now verify the four criteria of Prop. 5.46.

1. Equivalent to fullness condition (1).
2. Equivalent to fullness condition (2).
3. Let $A = [p, q]$ be in Φ, and $B = [0, 1]$ be in Ψ. If $\mathcal{H}(A \cap B) = \mathcal{H}([p, q]) = q - p = 0$, then A contains a single state $p = q$. Hence, $\#(A \cap B) = 1$.

4. Let us choose $A = [p, q] \in \Phi$ and $B = [0, 1] \in \Psi$. Then, $\forall j \in \Sigma$,

$$
\begin{aligned}
\mathcal{H}(f_1(A) \cap f_j^{-1}(B)) &= \mathcal{H}([\frac{p}{3}, \frac{q}{3}]) \\
&= \frac{1}{3}(q - p) = \frac{1}{3} \cdot \mathcal{H}(A \cap B) \\
\mathcal{H}(f_2(A) \cap f_j^{-1}(B)) &= \mathcal{H}([\frac{2+p}{3}, \frac{2+q}{3}]) \\
&= \frac{1}{3}(q - p) = \frac{1}{3} \cdot \mathcal{H}(A \cap B).
\end{aligned}
$$

Hence, there exists a $k < 1$ (here equal to $\frac{1}{3}$) such that, $\forall i, j \in \Sigma$,

$$
\mathcal{H}(f_i(A) \cap f_j^{-1}(B)) \leq k \cdot \mathcal{H}(A \cap B).
$$

This proves atomicity of the global invariant of this example.

The Cantor-set relation is very similar to the inverse of the dyadic map. Since our criteria for proving fullness and atomicity are all symmetric in time, the analyses of both systems look pretty much the same. Let us close the section with a last example: the logistic map $f(x) = \lambda x(1 - x)$. Apart from some more "smoothness", this map shows the same behavior again. In fact, for some values of the parameter λ, it is possible to prove that the system has a full and atomic invariant set using the same criteria again.

Example 5.50 (Logistic map). To fix the ideas, we concentrate on (see Fig. 5.4)

$$
f(x) = 5x(1 - x).
$$

This function is 2-to-1, and can be rewritten as a union of two injective branches:

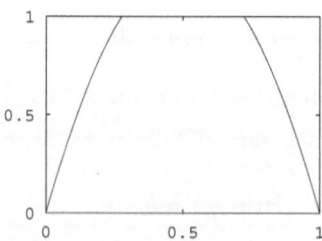

Fig. 5.4. Logistic map $f(x) = 5x(1 - x)$ on $[0, 1]$

$$
\begin{aligned}
f_1 &= ([0, \frac{1}{2}] \rightarrow 5x(1 - x)) \\
f_2 &= ([\frac{1}{2}, 1] \rightarrow 5x(1 - x)) \\
f &= f_1 \cup f_2.
\end{aligned}
$$

To help our subsequent development, let us evaluate the two inverse branches of f: the two roots of $f(x) = y$ are

$$\frac{1}{2}(1 \pm (1 - \frac{4}{5}y)^{\frac{1}{2}}).$$

The approximating sets for fullness and atomicity are (as in Ex. 5.48):

$$
\begin{aligned}
\varPhi &= \{[0,1]\} \\
\varPsi &= \{[p,q] \mid 0 \le p \le q \le 1\}
\end{aligned}
$$

and the decreasing function is:

$$\mathcal{H}([p,q]) = q - p.$$

Fullness. To apply Prop. 5.43, we verify three criteria.

1. Each $J_{i,}$ is obviously in $\varPhi$:

$$J_{i,} = f_i([0,1]) = [0,1]$$

and each $J_{,j}$ is in $\varPsi$:

$$
\begin{aligned}
J_{,1} &= f_1^{-1}([0,1]) = [0, \frac{1}{2}(1 - (\frac{1}{5})^{\frac{1}{2}})] \\
J_{,2} &= f_2^{-1}([0,1]) = [\frac{1}{2}(1 + (\frac{1}{5})^{\frac{1}{2}}), 1].
\end{aligned}
$$

2. For any $A \in \varPhi$ (there is only one such A), each $f_i(A) \in \varPhi$:

$$f_i(A) = f_i([0,1]) = [0,1]$$

and for any $B = [p,q] \in \varPsi$, each $f_j^{-1}(B) \in \varPsi$:

$$
\begin{aligned}
f_1^{-1}([p,q]) &= [\frac{1}{2}(1 - (\frac{4}{5}p)^{\frac{1}{2}}), \frac{1}{2}(1 - (\frac{4}{5}q)^{\frac{1}{2}})] \\
f_2^{-1}([p,q]) &= [\frac{1}{2}(1 + (\frac{4}{5}q)^{\frac{1}{2}}), \frac{1}{2}(1 + (\frac{4}{5}p)^{\frac{1}{2}})].
\end{aligned}
$$

3. Finally, any intersection of a set $A \in \varPhi$ with a set $B \in \varPsi$ is nonempty:

$$\forall 0 \le p \le q \le 1, [0,1] \cap [p,q] = [p,q] \ne \emptyset.$$

From this, we conclude the the global invariant of this system is full.

Atomicity. Let us now verify the four criteria of Prop. 5.46.

1. Equivalent to fullness condition (1).
2. Equivalent to fullness condition (2).
3. Let $A = [0, 1]$ be in Φ, and $B = [p, q]$ be in Ψ. If $\mathcal{H}(A \cap B) = \mathcal{H}([p, q]) = q - p = 0$, then A contains a single state $p = q$. Hence, $\#(A \cap B) = 1$.
4. Let us choose $A = [0, 1] \in \Phi$ and $B = [p, q] \in \Psi$. Then, $\forall i, j \in \Sigma$,

$$
\begin{aligned}
\mathcal{H}(f_i(A) \cap f_j^{-1}(B)) &= \frac{1}{2}((1 - \frac{4}{5}p)^{\frac{1}{2}} - (1 - \frac{4}{5}q)^{\frac{1}{2}}) \\
&= \frac{1}{2} \frac{(1 - \frac{4}{5}p) - (1 - \frac{4}{5}q)}{(1 - \frac{4}{5}p)^{\frac{1}{2}} + (1 - \frac{4}{5}q)^{\frac{1}{2}}} \\
&= \frac{2}{5} \frac{q - p}{(1 - \frac{4}{5}p)^{\frac{1}{2}} + (1 - \frac{4}{5}q)^{\frac{1}{2}}}.
\end{aligned}
$$

Since

$$
(1 - \frac{4}{5}p)^{\frac{1}{2}} \geq 5^{-\frac{1}{2}}
$$

we have an upper bound

$$
5^{-\frac{1}{2}}(q - p).
$$

Hence, there exists a $k < 1$ (here equal to $5^{-\frac{1}{2}}$) such that, $\forall i, j \in \Sigma$,

$$
\mathcal{H}(f_i(A) \cap f_j^{-1}(B)) \leq k \cdot \mathcal{H}(A \cap B).
$$

This proves atomicity of the global invariant of this example.

5.4 Attraction

The second important dynamical property investigated in this chapter is attraction. Roughly speaking, a set P is attracted to a set Q by evolution of f if it possible to go from P to Q in a certain number of (forward) iterations of f [326, 9]. In control theory and program semantics, attraction is often particularized to finite-time reachability or termination [187, 327, 91, 93]. As for invariance, the weaker notion of potential reachability exists in temporal logic, and permits to express properties of transition systems [98, 19].

In this section, we first give some intuitive aspects of attraction, starting from the presentation of reachability in §4.5. Then, using observation traces as in the characterization of invariant structures, we formalize attraction, and we propose a general taxonomy of attraction properties.

Our presentation of these classical concepts is partially based on the technical presentation of [116, 290].

5.4.1 Intuition: From Reachability to Attraction

In the first example, attraction happens when iterating a system with an asymptotically attracting fixed point: the system

$$f_1(x) = ([0,1] \rightarrow \frac{1}{3}(x+1) \leftarrow [\frac{1}{3}, \frac{2}{3}])$$

attracts $[0,1]$ to $\frac{1}{2}$ (see Fig. 5.5).

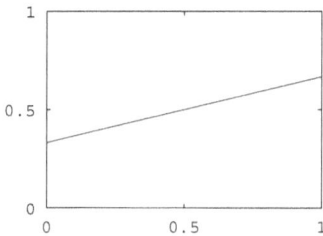

Fig. 5.5. Attracting fixpoint function $\frac{x+1}{3}$

The second example

$$f_2(x) = ([0, \frac{2}{3}] \rightarrow x + \frac{1}{3} \leftarrow [\frac{1}{3}, 1]) \cup ([\frac{1}{2}, 1] \rightarrow x \leftarrow [\frac{1}{2}, 1])$$

shows the attraction to a set of states, the interval $[\frac{1}{2}, 1]$ (see Fig. 5.6). Here,

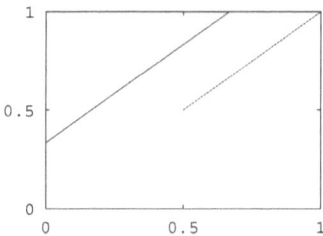

Fig. 5.6. Attraction to a set of states, $[\frac{1}{2}, 1]$

the attracting set is reached after a finite amount of time.

Again, we examine the qualitative property presented in §4.5. Reachability of a set Q reads (see Def. 4.24):

$$\mathbb{E}(Q, S) \equiv \forall s \in S, \exists n, s_n \in Q.$$

Thus, Q is reachable if every history eventually passes through Q, in finite time.

The set of histories can be $\theta(X, f)$, (X, f) being a RDS, or any set of histories starting from states of a particular set P. Then, similarly to invariance, we have two notions:

– Q is *necessarily reachable* from P iff

$$\forall x \in P, \forall s \in \theta(x, f), \mathbb{E}(Q, s);$$

– Q is *potentially reachable* from P iff

$$\forall x \in P, \exists s \in \theta(x, f), \mathbb{E}(Q, s).$$

These two versions both define a finite-time reachability. In order to get different reachability cases, we could

– relax the finite-time assumption, e.g.

$$\forall x \in P, \forall s \in \theta(x, f), \mathbb{E}^\omega(Q, s)$$

where $\mathbb{E}^\omega(Q, s) \equiv (s_\omega \in Q)$;
– precise the rest of an history after meeting Q;
– relax the first quantifier to get a partial reachability, e.g.

$$\exists x \in P, \forall s \in \theta(x, f), \mathbb{E}(Q, s).$$

Hereafter, we investigate the first possibility, namely infinite-time reachability, which can be seen as a limit of the finite-time case. The second case will be treated in §5.6, as a combination of attraction and invariance: indeed, if Q is invariant, we have a precision on what comes after Q (in this case, Q simply comes after Q).

5.4.2 From Weak to Full Attraction

Weak Attraction. Attraction is to termination what infinite iterations are to finite ones. A system f beginning in a set P terminates in a set Q if, for each initial state in P, f necessarily terminates after a finite number of iterations and must then reach a state in Q. A set Q attracts a set P by a system f if, after each realizable infinite iteration beginning in P, the resulting state belongs to Q, i.e. $f_\omega(P) \subseteq Q$.

Definition 5.51 (Weak attraction). *System f weakly attracts P to Q iff*

$$\forall \sigma \in \Sigma^\omega, f_\sigma(P) \subseteq Q.$$

Definition 5.52 (Attraction basin). *The attraction basin of a set Q by a system f is the largest set that is attracted by Q when iterating f.*

We can compare this concept of weak attraction with partial correctness of programs: if the program terminates, then it is correct. Here, whenever some infinite future σ exists, then P is attracted by Q when iterating f according to σ. The problem with this definition is that nothing prevents the case where there is no realizable infinite future, viz. $f_\sigma(P)$ is always $\emptyset$. However, we present it first because it gives the essence of the phenomenon of attraction. We now give three other possible definitions of attraction; each of them adds a condition avoiding the future of P to be empty.

Simple Attraction. We speak of "simple attraction" when at least one state of P has a potential infinite future.

Definition 5.53 (Simple attraction). *System f simply attracts P to Q iff*

$$(\forall \sigma \in \Sigma^\omega, f_\sigma(P) \subseteq Q) \wedge (P \cap J_+ \neq \emptyset).$$

So we are sure that $f_\sigma(P) \neq \emptyset$ for some σ, viz. $f^\omega(P) \neq \emptyset$.

Strict Attraction. We speak of "strict attraction" when P is not empty and every state of P has a potential infinite future.

Definition 5.54 (Strict attraction). *System f strictly attracts P to Q iff*

$$(\forall \sigma \in \Sigma^\omega, f_\sigma(P) \subseteq Q) \wedge (P \subseteq J_+) \wedge (P \neq \emptyset).$$

Example 5.55. The system defined by (see Fig. 5.7)

$$f(x) = ([0,1] \rightarrow \frac{x}{2} \leftarrow [0, \frac{1}{2}]) \cup ([0,1] \rightarrow \frac{x}{3} \leftarrow [0, \frac{1}{3}])$$

strictly attracts $[0,1]$ to the point $(x = 0)$.

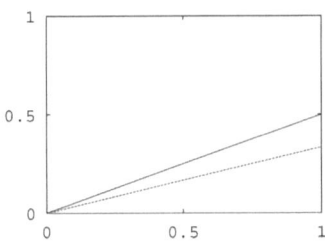

Fig. 5.7. Relation $\frac{x}{2} \cup \frac{x}{3}$ on $[0,1]$

Full Attraction. Finally, we speak of "full attraction" when, for each possible trace, there exists at least one state of P with that trace as potential infinite future.

Definition 5.56 (Full attraction). *System f fully attracts P to Q iff*

$$\forall \sigma \in \Sigma^\omega, (f_\sigma(P) \subseteq Q) \wedge (f_\sigma(P) \neq \emptyset).$$

We call it full attraction because the second part of the conjunction looks like the definition of fullness of the negative invariant J_-:

$$\forall \sigma \in \Sigma^\omega, f_\sigma(X) \neq \emptyset$$

viz.

$$\forall \sigma \in \Sigma^\omega, J_{\sigma,} \neq \emptyset.$$

Summary and Comments. In summary, we have the following proposition, the proof of which is left to the reader.

Proposition 5.57. *Attraction properties are ordered as follows:*

$$\left. \begin{array}{l} \textit{full attraction} \\ \textit{strict attraction} \end{array} \right\} \Rightarrow \textit{simple attraction} \Rightarrow \textit{weak attraction}.$$

Remark 5.58. – It is clear that full attraction and strict attraction are based on complementary conditions and are thus unrelated. It could be different if we had defined full attraction as:

$$\forall \sigma \in \Sigma^\omega, (f_\sigma(P) \subseteq Q) \wedge (f_\sigma(P) \neq \emptyset) \wedge (P \subseteq J_+) \wedge (P \neq \emptyset).$$

which obviously implies strict attraction.
– If an undefined state $\heartsuit$ is added, every P has an infinite future. In this case, it is convenient to add an assumption guaranteeing that the future of P contains more than just $\heartsuit$. This amounts to slightly modifying Def(s). 5.51–5.56.
– It is clear that termination on Q can be subsumed under attraction, by adding a "silent" transition $(Q \rightarrow id \leftarrow Q)$ to the considered system. It has the same function as the "idling" transition in elementary transition systems, or the *skip* instruction in Dijkstra's guarded command language, viz. absence of effect [91, 210].
– Even in the case of a well-defined ω-time reachability, there is no mention of what comes after Q: the system can stay in Q or it can iterate toward another set Q' that will be reached after another infinite-time for example.

5.4.3 A Taxonomy of Attraction

We have presented four variants of attraction, starting from the essential phenomenon called "weak attraction". In this section, our aim is to present variants more exhaustively and to relate some of them to the previous ones.

Let P and Q be two subsets of X, and f a relation on X. Our concern, "P attracted to Q", is summarized by the following questions.

– What length do we want the future of P to be?
– Does the future of points of P lead to Q? Necessarily or potentially?
– Do points of P have a long enough future?

More formally, these questions give the following, where $x \in P$ and $\sigma \in (\mathbb{P}(\Sigma))^\alpha$.

– Traces of transitions belong to $(\mathbb{P}(\Sigma))^\alpha$ with

$$\alpha \in \{\leq n, n, *, \omega, \infty, \mathbb{O}\}.$$

– Do we have a necessary or potentially attraction to Q, respectively (see Def. 5.4):

$$f_\sigma(\{x\}) \subseteq Q$$

or

$$\{x\} \subseteq f_\sigma(Q) \ ?$$

– Does P have a long enough future,

$$f_\sigma(\{x\}) \neq \emptyset \ ?$$

Let us define

$$
\begin{aligned}
P_1(x,\sigma) &= f_\sigma(\{x\}) \subseteq Q, \\
P_2(x,\sigma) &= \{x\} \subseteq f_\sigma(Q), \\
P_3(x,\sigma) &= f_\sigma(\{x\}) \neq \emptyset.
\end{aligned}
$$

There are six possible ways of quantifying traces of $(\mathbb{P}(\Sigma))^\alpha$ and points of P:

$$\mathcal{Q} = \{\forall x \forall \sigma, \forall x \exists \sigma, \exists x \forall \sigma, \exists x \exists \sigma, \forall \sigma \exists x, \exists \sigma \forall x\}.$$

Basic attraction properties are thus, $\forall a, b \in \mathcal{Q}, \forall P_4 \in \{P_1, P_2\}$:

$$a P_4,$$
$$a P_4 \wedge b P_3,$$
$$a(P_4 \wedge P_3),$$

and their conjunctions.

Of course, all these possibilities can be compared w.r.t. implication. For instance, $a P_4 \wedge b P_3 \Rightarrow a P_4$, and $a(P_4 \wedge P_3) \Rightarrow a P_4 \wedge a P_3 \Rightarrow a P_4$. An order is established between quantifyers of the same property:

$$\forall \sigma \forall x \Rightarrow \left\{ \begin{array}{l} \exists x \forall \sigma \\ \exists \sigma \forall x \\ \forall x \exists \sigma \\ \forall \sigma \exists x \end{array} \right\} \Rightarrow \exists \sigma \exists x,$$

and between their conjunctions.

Let us now give the four cases presented in §5.4 in the light of our formalization, with $\alpha = \omega$.

– Weak attraction: $(\forall\sigma\forall x, f_\sigma(\{x\}) \subseteq Q)$.
– Simple attraction: weak $\wedge$ $(\exists x\exists\sigma, f_\sigma(\{x\}) \neq \emptyset)$.
– Strict attraction: weak $\wedge$ $(\forall x\exists\sigma, f_\sigma(\{x\}) \neq \emptyset)$.
– Full attraction: weak $\wedge$ $(\forall\sigma\exists x, f_\sigma(\{x\}) \neq \emptyset)$.

Notation 5.59. *In the following, we denote simple attraction from P to Q by a system f as follows:*

$$P \overset{f}{\leadsto} Q$$

which means $\emptyset \neq f^\omega(P) \subseteq Q$.

5.5 Attraction Criteria

In this section, we present sufficient conditions to prove that a set Q attracts a set P by a system f, as we did for fullness and atomicity. The *Lyapunov-like criteria* given here generalize the ones developed in [116, 290]. As in atomicity criteria, we use a decreasing function related to the "distance" toward the attracting set.

The following proposition focuses on strict attraction; other variants could be proved using the same reasoning.

Proposition 5.60 (Attraction criteria). *To prove strict attraction of a RDS (X, f) observed on a covering $\alpha = \{f_i \mid i \in \Sigma\}$, it suffices to find a family Ψ of nonempty sets and a function $\mathcal{H}$ from sets of X to $\mathbb{R}$ such that:*

1. *$P \in \Psi$;*
2. *$\forall i \in \Sigma, A \in \Psi, f_i(A) \in \Psi$;*
3. *$\forall A \in \Psi, (\mathcal{H}(A) = 0) \Rightarrow (A \subseteq Q)$;*
4. *$\exists k, 0 \leq k < 1, \forall i \in \Sigma, \forall A \in \Psi, \mathcal{H}(f_i(A)) \leq k \cdot \mathcal{H}(A)$;*
5. *$P \neq \emptyset$ and $P \subseteq J_+$.*

Proof. It is easy to prove by induction:

$$\forall n \geq 0, \forall \sigma \in \Sigma^n, f_\sigma(P) \in \Psi.$$

The basic case is given by (1) and the induction is based on (2).

Using (4), we can then prove:

$$\forall n \geq 0, \forall \sigma \in \Sigma^n, \mathcal{H}(f_\sigma(P)) \leq k^n \mathcal{H}(P)$$

and thus

$$\forall \sigma \in \Sigma^\omega, \mathcal{H}(f_\sigma(P)) = 0.$$

Finally, using (3) gives

$$\forall \sigma \in \Sigma^\omega, f_\sigma(P) \subseteq Q.$$

which, using (5), is equivalent to the definition of strict attraction.

Remark 5.61. – It is possible to give more refined criteria to prove attraction [116]. They are all based on the two following central keys:
1. approximation of P and its successive iterations,
2. definition of a decreasing function $\mathcal{H}$.
– This technique seems very powerful to prove attraction; its crucial point is most of the time the discovery of an adequate decreasing function.
   When the system is contracting for some metric, or $P = X$ and monotonicity entails a decreasing sequence of iterates, exhibiting a decreasing function is not too difficult: the diameter of successive iterates $f^n(X)$ often suffices.
   In other cases, e.g. for arbitrary P and Q, proving attraction using these criteria is not so easy. The decreasing function can be very tricky.

5.6 Attraction by Invariants

In this section, we examine how attraction and invariance can be combined, as this case is very often treated in the literature: the attracting set is invariant. Actually, this is a particular case of precising what happens after having reached the attracting set.

We have seen that an invariant and its structure are fundamental for understanding the dynamics of a system. On the other hand, we have formalized the notion of attraction, which, roughly speaking, ensures to go from a set P to a set Q after a certain number of iterations. Here, we try to see how the system attracts a part of the space to its invariant. Let us start with a simple example.

Example 5.62 (Cantor relation, cont'd). Consider $f_1(x) = \frac{x}{3}$ again. It is easy to see that $[0, 1] \stackrel{f_1}{\rightsquigarrow} \{0\}$.

If we add the second function $f_2(x) = \frac{x}{3} + \frac{2}{3}$ and consider their union $f_1 \cup f_2$, we obtain the Cantor relation. Its invariant J is the middle-thirds Cantor-set in $[0, 1]$; it is full and atomic. We see that $[0, 1]$ is attracted to this invariant. Let us prove J attracts the domain $P = [0, 1]$ of f. In this system, we have $J = J_-$ and $P = J_+$.

Proposition 5.63. *Let* (X, f) *be a RDS. Then,*

$$J_- \subseteq J_+ \Rightarrow J_+ \stackrel{f}{\rightsquigarrow} J.$$

Proof.

$$J_- = \cup_{\sigma \in \Sigma^\omega} f_\sigma(X)$$
$$\Rightarrow \quad \forall \sigma \in \Sigma^\omega, f_\sigma(X) \subseteq J_-$$
$$\Rightarrow \quad X \stackrel{f}{\rightsquigarrow} J_-$$

$$\Rightarrow \quad J_+ \overset{f}{\rightsquigarrow} J_-$$
$$\because \ \text{hyp.} \ \Rightarrow J = J_- \cap J_+ = J_-$$
$$\Rightarrow \quad J_+ \overset{f}{\rightsquigarrow} J.$$

Example 5.64 (Cantor relation, Ex. 5.62 revisited). Moreover, we also have strict attraction of P to $J_- = J$ because $P \subseteq J_+$ (in this case, we have $P = J_+$) and the positive invariant is not empty.

The inverse of the Cantor map has the same Cantor set as invariant. Yet, this invariant does not anymore attract the domain $[0, 1]$: the inverse Cantor map is repulsing.

In fact, the invariant of the direct Cantor map equals its negative invariant, whereas the invariant of the inverse Cantor map equals its positive invariant. We have noticed that a negative invariant results from infinite traces, and a positive invariant begins infinite traces (see §5.1).

Example 5.65. Let f be the following system (see Fig. 5.8):

$$
\begin{aligned}
f(x) \ = \ & [0,\tfrac{1}{2}]\times[0,\tfrac{1}{2}]\to(2x) \quad \times(\tfrac{y}{4}) \quad \leftarrow[0,1]\times[0,\tfrac{1}{8}] \\
\cup \ & [\tfrac{1}{2},1]\times[0,\tfrac{1}{2}]\to(2x-1)\times(\tfrac{y}{4}+\tfrac{4}{8})\leftarrow[0,1]\times[\tfrac{4}{8},\tfrac{5}{8}] \\
\cup \ & [0,\tfrac{1}{2}]\times[\tfrac{1}{2},1]\to(2x) \quad \times(\tfrac{y}{4}+\tfrac{2}{8})\leftarrow[0,1]\times[\tfrac{3}{8},\tfrac{4}{8}] \\
\cup \ & [\tfrac{1}{2},1]\times[\tfrac{1}{2},1]\to(2x-1)\times(\tfrac{y}{4}+\tfrac{6}{8})\leftarrow[0,1]\times[\tfrac{7}{8},1].
\end{aligned}
$$

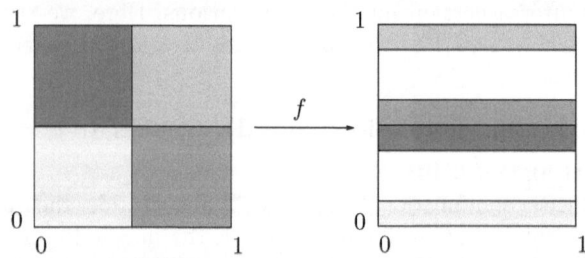

Fig. 5.8. Effect of f on $[0,1]^2$

The positive invariant J_+ is by definition stable under f^{-1}, viz. preserved in the future, and it includes the full square $[0,1] \times [0,1]$. Thus it includes J_-, generated by f and preserved in the past; J_- consists of horizontal segments of length one. Since $J_- \subseteq J_+$, we have $J = J_-$. The invariant J attracts the square $[0,1] \times [0,1]$ since J_- attracts it.

If we modify the above system so that f^{-1} is contracting along x by 4 instead of 2, the positive invariant J_+ may become a strict subset of the full square, viz. a set of scattered vertical segments. Then, J_- is not included in J_+ anymore, and we cannot prove, using the present approach, that J attracts the whole square. We should then use the criteria developed in §5.5.

Besides proving a full and atomic invariant is attracting, we may wish to prove that a given attracting set generates rich dynamics, i.e. is strongly dependent on initial conditions and topologically transitive. To verify this, one could use variants of the criteria for detecting fullness and atomicity (see §§5.2, 5.3). Again, the problem is to discover adequate families of approximating sets and an adequate convergence function. This is left for future work.

5.7 Discussion

In this last section, we first present notions related to invariance, attraction, fullness, and atomicity; energy-like functions are then compared to decreasing functions used in atomicity and attraction criteria; finally, we close the chapter by giving an informal view of dynamical complexity using the tools introduced up to now.

5.7.1 Invariance and Attraction: Related Notions

Necessary and potential invariance as well as attraction are classical in dynamical systems and program theory, as well as in temporal logic [9, 91, 98, 196, 326].

Invariance. Necessary (resp. potential) invariants are equivalent to Sifakis' invariants (resp. non-terminating trajectories) [284]. Hutchinson's invariants are related to potential invariants [159, 140, 28, 325], though they are obtained by successive iterations using the contraction mapping theorem, which in this case gives the same limit as lattice fixpoint theorems. Maximum invariant sets as defined in [305] are also related to potential invariants.

Invariants of programs are properties that are verified by the variables of these programs during their evolutions. Basically, invariants are associated to iterations (cf. DO-loops in Dijkstra's nondeterministic guarded command language [91, 93]): they must be verified all along the computations of loops, in order to preserve partial correctness.

The temporal logic notion of *safety* is related to invariance. Informally, "nothing bad" happens to the system involved, that is, safety properties concern states or transitions that are always verified, for all histories of the system [192]. Safety is thus very close to Def. 4.23. The following notation is used to say that a property P is invariant: $\Box P$. See also [193, 210, 2, 67, 81] for more on temporal logic, and fixed point calculus for temporal logic.

Necessary/potential invariance can be defined using two binary operators $A[\cdot \cup \cdot]$ and $E[\cdot \cup \cdot]$ of Computation Tree Logic (CTL) [99, 98, 19]:

$$\text{necessary invariance} : J \models \neg E[1 \cup \neg J]$$
$$\text{potential invariance} : J \models \neg A[1 \cup \neg J].$$

In control theory, controllability is equivalent to finding an invariant of the system, a subspace in which the system acts as predicted [18, 327, 129].

Attraction. Total correctness of programs is obtained by conjugation with a Lyapunov-like technique: a decreasing function has to be found, such that its minimum is reached after a finite-time.

In Unity, attraction also plays an important role, under the construction "P leadsto Q", equivalent to a finite-time version of $P \rightsquigarrow Q$ [61].

The temporal logic notion of *liveness* is related to attraction. Informally described in [192] and formally defined in [15], liveness means that "something good" is supposed to happen to the system. The notation used, $\Diamond P$, means that eventually, property P must be verified, whatever the history of the system is. This property can be a predicate defined on states or transitions. Remark that liveness is very close to Def. 4.24, and $\Diamond$ and $\Box$ are duals: $\neg \Diamond \neg = \Box$.

Necessary/potential finite-time attraction can be defined using the same two binary operators $A[\cdot \cup \cdot]$ and $E[\cdot \cup \cdot]$ of CTL:

$$\text{necessary reachability} : P \models E[1 \cup Q]$$
$$\text{potential reachability} : P \models A[1 \cup Q].$$

In control theory, finite-time attraction is called *reachability* [18, 327, 129]. In the abstract framework of semigroups, it is known as *absorption* [191].

Fullness and Atomicity. Fullness and atomicity have been introduced in [286]: they are respectively equivalent to exactness and weak generation [196]. Indeed, a dynamical system (X, f) is said to be *exact* when $\lim_{n \to \omega} \mu(f^n(A)) = 1$ for every $A \subseteq X$ such that $\mu(A) > 0$, where μ is a measure on X. A finite open cover $\alpha = \{A_i\}$ of X is a *weak generator* for f if for all sequence $(A_{i_k})_k \in \alpha^{\mathbb{Z}}$, the intersection $\cap_k f^k(A_{i_k})$ contains no more than one point.

Other authors have proposed to study the structure of invariants using symbolic observations of evolutions of systems. We have already cited symbolic dynamics (see §4.7.2), but more specifically, formal language theory is sometimes used to describe the inner structure of invariants [72, 139, 170, 171, 307, 308, 322, 334].

Fairness. Let us thus examine a last notion from temporal logic. Fairness [110, 61, 19] is built on both invariance and attraction. Indeed, fairness properties can be obtained by composing several basic blocks of the form $\Box \Diamond P$, using logical connectors (negation, disjunction, conjunction). For example, P stating a property on states of a system, and R precising a property on some transitions of the system, the following lines express standard fairness properties:

$$\text{(weak)} \qquad \Diamond \Box P \Rightarrow \Box \Diamond R$$
$$\text{(strong)} \qquad \Box \Diamond P \Rightarrow \Box \Diamond R.$$

5.7.2 Energy-Like Functions

We used energy-like functions to develop criteria for atomicity and attraction: Lyapunov-like or Floyd-like functions are used, depending on the domain structure. We review some well-known examples where such functions play an important role to prove convergence.

Classical Dynamical Systems. In the context of classical dynamical systems, a fixed point is already an invariant. An attracting fixed point is of course an attractor. Lyapunov functions are used to prove the convergence of points to the fixed point [88, 326, 286, 9, 196, 221].

Neural Networks. Neural networks are used as associative memories. Each stored pattern corresponds to a fixed point or a periodic orbit of the configuration space, and the dynamics of the system has to converge to this attractor as fast as possible, and from a large enough neighborhood. In this context, energy-like functions are used to prove stabilization; they also permit to build optimizing systems [60, 119, 120, 121, 142, 173, 212, 211].

Program Theory. Programs are also studied through invariance and attraction. A loop is often characterized by an invariant predicate. Convergence here is finite, and a termination function is used in the same way as energy functions or Lyapunov functions [91, 93, 109]. When non-halting programs are needed, like in real-time control systems, invariance is used to denote "normal" states. When a problem occurs, the system has to enter in a self-stabilizing mode, viz. a transient dynamics leading back to the invariant, which is thus attracting [61, 90, 91, 93, 123, 278, 327].

5.7.3 Dynamical Complexity

In nonlinear dynamical systems, papers and books on dynamical complexity are numerous. For instance, see [22, 88, 154, 196, 217, 220, 230, 326] and reference therein, and [25, 33, 177, 176, 270, 331] more specifically concerning high-dimensional systems. However, the intrinsic difficulty of many interesting complex dynamical phenomena seems to resist against any precise definition of what dynamical complexity is. Despite this difficulty, one of the main motivations of this monograph is the understanding of some complex phenomena arising in discrete-time dynamical systems. Although we do not answer to the principal question of defining complexity, we hope that our contribution constitutes a step toward this objective that we share with many others. Let us review some important directions to characterize complexity.

- Of course, there is "chaos" [9, 88, 154, 181, 326] and related mathematical definitions based on structural and topological properties.
  Chaos does not perfectly fit all of complex behaviors, especially when high-dimensional systems are considered. In Chap. 8, we will come back on this problem as a motivation to propose a new classification of cellular automata behaviors.
- Ergodic theory and measure theory are useful to characterize some complex behaviors [21, 196, 285]. They offer powerful mathematical tools to evaluate the disorder introduced in the state space by the evolution of some systems. Again, the results are not always transferable to all kinds of systems.

– Measures from physics and statistical mechanics can be used to evaluate the complexity of systems, generally by experiments or simulations. For example, entropy, invariant measures, Lyapunov exponents, fractal dimension of attractors, which all indicate some complexity [217, 33, 196].

– Finally, lots of interesting results involving number theory have been published on the complexity of symbolic sequences. These results can also be related to symbolic dynamics and coding theory. Among others, see [11, 12, 202, 268].

Our view is the following: complexity measures from physics provide a suitable definition of complex behaviors but they often rely on experiments and do not always result from analytical computations. The idea is to take "the best of two worlds" (S. Getz): to go as far as possible using analytical developments that remain compatible with intuition and observation, and to switch to physical measures to complete the analysis.

We have shown that complex behaviors of a system are closely related to the observed structure of its invariant. Fullness entails chaos at the level of symbolic traces (Prop(s). 5.32 and 5.35). Atomicity restricts the amount of states corresponding to these histories, which leads to topological transitivity and sensitivity to initial and final conditions (Prop(s). 5.38 and 5.39). Fullness and atomicity do not entail density of periodic points, but only density of periodic traces. However, they are sufficient to imply Knudsen chaos.

Attraction can also give information on the complexity of systems. For example, the size and the number of attractors is important. In §5.6, we have related invariance and attraction in a unique phenomenon: attraction by invariants, which reinforces this view. Indeed, in this case, the invariant structure also characterizes the attractor using fullness and atomicity.

From this, a natural measure of complexity consists in analyzing the internal structure of attractors [171]. It is clear for example, that fixpoints offer simpler behaviors than periodic cycles, which in turn appear less complex than quasi-periodic orbits, and aperiodic ones. In [9], the author builds a hierarchy of periodic behaviors of dynamical systems, from strict fixpoints to chain-recurrent points, as fixed (sets of) points of different relations. In [25] a similar hierarchy of recurrent behaviors is also presented, as complexity hierarchy of dynamical systems. Fairness (and its many variants, see [110]) can also be introduced between quasi-periodicity and aperiodicity because it entails a recurrence on states without precision on the amount of time needed to come back to a given state.

Based on these ideas, we propose in Chap. 8 an attraction-based classification of cellular automata behaviors that is used as complexity measure.

6. Compositional Analysis of Dynamical Properties

In this chapter, we systematically analyze dynamical properties, namely invariance and attraction, of composed systems in terms of similar individual properties of their components.

Dynamical systems were defined in Chap. 2, and we showed how to structure them with the help of composition operators introduced in Chap. 3. Abstract observation (see Chap. 4) was used to define invariance and attraction of relational systems in general (Chap. 5).

We now carry out the analysis of composed dynamical properties using the principle of compositional analysis: we exhibit interesting phenomena where complexity arises from composition of very simple components, particularly using union. This structural composition is of fundamental importance to understand the complexity of some well-known systems, as illustrated in Chap(s). 7 and 8.

The chapter is organized as follows. In §6.1, the objectives and results of the chapter are introduced; §§6.2–6.4 focus on unary operators: inversion, restrictions, negation; in §§6.5–6.8, we analyze sequential composition, intersection, union quite thoroughly, free product and we discuss connected product; in §6.9, we combine union and free product; finally, in §6.10, we close the chapter with a discussion.

6.1 Aims and Informal Results

Compositional Analysis. The goal of compositional analysis is the following. We want to analyze some qualitative dynamical property G of a dynamical system S. If this system happens to be decomposable into subsystems $S_1, \cdots, S_n$ using an operator introduced in Chap. 3, say $S = \star_i S_i$, its global analysis, viz. $G(S)$, could be reduced to an appropriate combination of individual analyses of its components, i.e. $\diamond_i I(S_i)$. This is summarized in a standard diagram:

F. Geurts: Abstract Compositional Analysis of Iterated Relations, LNCS 1426, pp. 135-161, 1998.
© Springer-Verlag Berlin Heidelberg 1998

$$S_i \xrightarrow{\quad I \quad} I(S_i)$$

$$\star \downarrow \qquad\qquad \downarrow \diamond$$

$$\star_i S_i \xrightarrow{\qquad\qquad} \cdot$$

$$G$$

Of course, to each system-composition operator "$\star$" corresponds a specific property-composition operator "$\diamond$".

This kind of development looks nicely algebraic. We use it to introduce the principle here, but we are not able to keep it as clear as this in the subsequent developments: all properties are not necessarily expressed in an algebraic way, at least not in such a simple form.

The goal expressed as above is probably too strong, only the $I(S_i)$'s are used to get the global property $G(S)$. In our case, G and I can return the invariant of system, information about its structure, like fullness or atomicity, its attraction properties, and so on. We could soften this requirement and admit information about the global system and all components in each $I(S_i)$, as far as the amount of information is lower than proving or computing $G(S)$ globally!

In what follows, we analyze these properties combined with the composition operators defined in Chap. 3. Difference is omitted simply because we study negation which does not bring interesting results. Connected product is just mentioned: its generality cannot be treated analytically. However, in Chap. 8, we will particularize connected product to cellular automata and, using complexity measures from physics, we will analyze a class of complex behaviors. In Chap. 9, the connected product will also be studied as a general model of classical computational models, and computational properties will be analyzed by composition.

Properties To Be Analyzed. As a basis for the next sections of this chapter, let us explain the developments we will carry out. Recall the invariants and the global attractor of a system are all computed by successive forward or backward iterations of the set-transformer corresponding to the relational dynamical system. Fullness and atomicity are directly attached to invariants. To set notations for the rest of this monograph, we will work with RDS (X, f) and (X, g). For the products, other systems will be mentioned.

At first, we will compute the "kind" of composed systems, restricting our attention to simple variant relations. This notion expresses the expanding, contracting or neutral type of variant relations; it is not very important as such but it can be used in the assumptions of some theorems.

Then, we will compute two invariants for each case of composition operator: the *greatest potential backward invariant*,

$$J_-^f = \cap_i f^i(X),$$

and the *least necessary forward invariant* greater than A

$$J_f^+(A) = \cup_i f^i(A),$$

with $A \subseteq f(A)$.

The whole space X is the top of the lattice $\mathbb{P}(X)$, and the decreasing iterations of a monotonic set-transformer lead to a non-trivial greatest fix-point. The bottom of $\mathbb{P}(X)$ is the empty set $\emptyset$ and successive iterations from it remain trivially equal to $\emptyset$, by excluded miracle. Thus, we have to chose $A \neq \emptyset$ that initiates an increasing sequence in order to get a non-trivial least fixpoint of f.

The greatest potential forward invariant J_+^f is obtained in a similar way to J_-^f, i.e. f is replaced by f^{-1}. This permits to get the greatest potential global invariant J^f.

Fullness and atomicity can then be examined. Given a composed system f, a covering α and its corresponding alphabet Σ, we want to determine the properties of J^f, the global potential invariant. Thus, we have to find a way to compute $f_i(A)$ for each $i \in \Sigma$ and $A \subseteq X$, because theses expressions are the basis of trace-parametrized invariants, fullness and atomicity.

Finally, *attraction* is investigated in the same way, whenever possible. Actually, we focus on the global attractor of the system f, which is J_-^f, the smallest Q such that $X \xrightarrow{f} Q$.

Informal Results. Among the unary operators, we examine inversion, restrictions and negation independently.

– Inversion does not introduce a real perturbation in the system if we consider duality between past and future, which means we always observe past and future together. Of course, attraction is treated differently because it is asymmetric: the inverse of attraction is repulsion.
– Restrictions preserve the properties of a system almost entirely but we still have to add some assumptions ensuring, for example, the invariant of a restricted system is equal to the restricted invariant of the system. In fact, depending on the restricted region of the domain or range of the system, complexity can decrease or can remain unchanged.
– Negation, on the contrary, gives a completely different system. This leads to a trivial potential invariant, whereas the necessary invariant has to be computed as such, without any possible short-cut.

With some n-ary operators, complexity can result from a composition, even when applied to originally simple systems.

– Sequential composition is not easy to study using only information on the components. It seems there is no possible short-cut, similarly to negation. Very often, the composed system has to be studied on its own. A decrease of complexity is possible.

- Intersection very much resembles restrictions, at least conceptually, but it can be much more irregular. The fundamental explanation resides in the fact that it is not always possible to express $f \cap g$ as a double (i.e. domain and range) restriction $(g_1 \to f \leftarrow g_2)$, which would come from $g = g_1 \times g_2$. Using intersection, complexity can decrease.
- Union is very interesting because it allows to add several independent dynamics and, thus, to generate more complexity than each independent system can. It encompasses many different kinds of behaviors, and deserves a deep investigation of its invariant structure.
- The free product is interesting because it allows to compose different spaces as well as different systems, which the previous operators cannot do. The resulting behavior can be seemingly more complex than the original ones but it remains easy to analyze: each component can be treated independently.
- This is not the case of the connected product, adding explicit interactions between components. This operator combines spaces, systems, and is of course able to increase complexity. The problem is that the analysis is generally quite difficult.

6.2 Inversion

This first case is very simple, and it allows us to propose a *duality principle*, w.r.t. time: all properties related to forward set-transformers of systems can be rephrased about backward versions of inverse systems, and vice versa.

Kind. Under some specific assumptions, if a system f is contracting (expanding), its inverse is expanding (contracting); if f is neutral, its inverse has the same kind.

Proposition 6.1. *Let (X, f) be a simple contracting RDS with contractivity factor $\gamma(f) < 1$, then*

$$\gamma(f^{-1}) = \frac{1}{\gamma(f)}$$
$$\kappa(f^{-1}) = -\kappa(f).$$

Proof. We have $\forall A, B \subseteq X$,

$$h(f(A), f(B)) \leq \gamma(f) \cdot h(A, B).$$

It is thus valid for $f^{-1}(A)$ and $f^{-1}(B)$, provided A and B both belong to $Dom(f^{-1}) = Rg(f)$:

$$h(f(f^{-1}(A)), f(f^{-1}(B))) \leq \gamma(f) \cdot h(f^{-1}(A), f^{-1}(B)).$$

As $A \subseteq Rg(f)$, $f(f^{-1}(A)) = A$.

and the *least necessary forward invariant* greater than A

$$J_f^+(A) = \cup_i f^i(A),$$

with $A \subseteq f(A)$.

The whole space X is the top of the lattice $\mathbb{P}(X)$, and the decreasing iterations of a monotonic set-transformer lead to a non-trivial greatest fix-point. The bottom of $\mathbb{P}(X)$ is the empty set $\emptyset$ and successive iterations from it remain trivially equal to $\emptyset$, by excluded miracle. Thus, we have to chose $A \neq \emptyset$ that initiates an increasing sequence in order to get a non-trivial least fixpoint of f.

The greatest potential forward invariant J_+^f is obtained in a similar way to J_-^f, i.e. f is replaced by f^{-1}. This permits to get the greatest potential global invariant J^f.

Fullness and atomicity can then be examined. Given a composed system f, a covering α and its corresponding alphabet Σ, we want to determine the properties of J^f, the global potential invariant. Thus, we have to find a way to compute $f_i(A)$ for each $i \in \Sigma$ and $A \subseteq X$, because theses expressions are the basis of trace-parametrized invariants, fullness and atomicity.

Finally, *attraction* is investigated in the same way, whenever possible. Actually, we focus on the global attractor of the system f, which is J_-^f, the smallest Q such that $X \overset{f}{\leadsto} Q$.

Informal Results. Among the unary operators, we examine inversion, restrictions and negation independently.

- Inversion does not introduce a real perturbation in the system if we consider duality between past and future, which means we always observe past and future together. Of course, attraction is treated differently because it is asymmetric: the inverse of attraction is repulsion.
- Restrictions preserve the properties of a system almost entirely but we still have to add some assumptions ensuring, for example, the invariant of a restricted system is equal to the restricted invariant of the system. In fact, depending on the restricted region of the domain or range of the system, complexity can decrease or can remain unchanged.
- Negation, on the contrary, gives a completely different system. This leads to a trivial potential invariant, whereas the necessary invariant has to be computed as such, without any possible short-cut.

With some n-ary operators, complexity can result from a composition, even when applied to originally simple systems.

- Sequential composition is not easy to study using only information on the components. It seems there is no possible short-cut, similarly to negation. Very often, the composed system has to be studied on its own. A decrease of complexity is possible.

- Intersection very much resembles restrictions, at least conceptually, but it can be much more irregular. The fundamental explanation resides in the fact that it is not always possible to express $f \cap g$ as a double (i.e. domain and range) restriction $(g_1 \rightarrow f \leftarrow g_2)$, which would come from $g = g_1 \times g_2$. Using intersection, complexity can decrease.
- Union is very interesting because it allows to add several independent dynamics and, thus, to generate more complexity than each independent system can. It encompasses many different kinds of behaviors, and deserves a deep investigation of its invariant structure.
- The free product is interesting because it allows to compose different spaces as well as different systems, which the previous operators cannot do. The resulting behavior can be seemingly more complex than the original ones but it remains easy to analyze: each component can be treated independently.
- This is not the case of the connected product, adding explicit interactions between components. This operator combines spaces, systems, and is of course able to increase complexity. The problem is that the analysis is generally quite difficult.

6.2 Inversion

This first case is very simple, and it allows us to propose a *duality principle*, w.r.t. time: all properties related to forward set-transformers of systems can be rephrased about backward versions of inverse systems, and vice versa.

Kind. Under some specific assumptions, if a system f is contracting (expanding), its inverse is expanding (contracting); if f is neutral, its inverse has the same kind.

Proposition 6.1. *Let (X, f) be a simple contracting RDS with contractivity factor $\gamma(f) < 1$, then*

$$\gamma(f^{-1}) = \frac{1}{\gamma(f)}$$
$$\kappa(f^{-1}) = -\kappa(f).$$

Proof. We have $\forall A, B \subseteq X$,

$$h(f(A), f(B)) \leq \gamma(f) \cdot h(A, B).$$

It is thus valid for $f^{-1}(A)$ and $f^{-1}(B)$, provided A and B both belong to $Dom(f^{-1}) = Rg(f)$:

$$h(f(f^{-1}(A)), f(f^{-1}(B))) \leq \gamma(f) \cdot h(f^{-1}(A), f^{-1}(B)).$$

As $A \subseteq Rg(f)$, $f(f^{-1}(A)) = A$.

Invariance.

Proposition 6.2. *The greatest potential backward invariant of f^{-1} is equal to the greatest potential forward invariant of f:*

$$J_-^{f^{-1}} = J_+^f.$$

Proof. Obvious:

$$J_-^{f^{-1}} = \cap_i (f^{-1})^i(X) = \cap_i f^{-i}(X) = J_+^f.$$

Combining Proposition 6.2 and its dual, we have the following corollary.

Corollary 6.3. *The greatest potential global invariants of a RDS (X, f) and its inverse (X, f^{-1}) are equal:*

$$J^{f^{-1}} = J^f.$$

In spite of its simplicity, this corollary is very interesting because it allows us to consider systems or their inverses in the same way: the results are independent from the direction of evolution. Of course, the same holds for necessary invariants.

Corollary 6.4. *The least necessary global invariants greater than A of a RDS (X, f) and its inverse (X, f^{-1}) are equal:*

$$J_{f^{-1}}^+(A) = J_f^-(A).$$

Structure of Invariants. Past and future traces are exchanged, which does not influence fullness and atomicity. The only technical addition is the introduction of an inverse covering.

Proposition 6.5. $J^{f^{-1}}$ *is full (resp. atomic) on a covering α iff J^f is full (resp. atomic) on the inverse covering α^{-1}.*

Proof. Given a covering $\alpha = \{A_i\}$ of f^{-1}, its alphabet Σ, and one of the possible transition i, let us compute the first parametrized forward iteration:

$$
\begin{aligned}
& f_i^{-1}(A) \\
= \ & (f^{-1} \cap A_i)(A) \\
= \ & (f \cap A_i{}^{-1})^{-1}(A) \\
= \ & f_i(A)
\end{aligned}
$$

where the last i stands for an index in the inverse covering $\alpha^{-1} = \{A_i{}^{-1} | A_i \in \alpha\}$.

Attraction. Attraction of the inverse of a relation gives rise to a new definition, namely attraction to the past or *repulsion*:

$$\emptyset \neq (f^{-1})^{\omega}(P) \subseteq Q.$$

Thus, the criteria proposed in §5.5 apply, provided we reverse the iteration direction.

6.3 Restrictions

6.3.1 Domain Restriction

Kind. In general, domain restrictions do not change the "kind" of variant systems:

$$\kappa(B \to f) \in \{\kappa(f), 0\}.$$

Invariance.

Proposition 6.6. *If (X, f) is an injective RDS, then the greatest potential backward invariant of its domain restriction $(B \to f)$ is*

$$J_{-}^{(B \to f)} = \cap_i f^i(B).$$

If, moreover, $f(B) \supseteq B$, then

$$J_{-}^{(B \to f)} = f(B).$$

Proof. By induction on n, we prove that $\forall n \geq 1, (B \to f)^n(X) = \cap_{i=1}^{n} f^i(B)$.

– Basic case $n = 1$: by Def. 3.3 and Prop. 3.30,

$$(B \to f)(X) = f(B \cap X) = f(B).$$

– Inductive case: by injectivity of f,

$$\begin{aligned}
(B \to f)^{n+1}(X) &= (B \to f)((B \to f)^n(X)) \\
&= (B \to f)(\cap_{i=1}^{n} f^i(B)) \\
&= f(B \cap \cap_{i=1}^{n} f^i(B)) \\
&= \cap_{i=1}^{n+1} f^i(B).
\end{aligned}$$

Thus, $J_{-}^{(B \to f)} = \cap_{n \geq 1} f^n(B)$. If moreover, $B \subseteq f(B)$, then $J_{-}^{(B \to f)}$ reduces to $f(B)$.

Proposition 6.7. *If (X, f) is an injective RDS, then the least potential forward invariant greater than A of its domain restriction $(B \to f)$ is*

$$J^+_{(B \to f)}(A) = \cup_{n \geq 1}(f^n(A) \cap \cap^n_{i=1} f^i(B)).$$

Moreover,

$$
J^+_{(B \to f)}(A) = \begin{array}{ll}
f(B) \cap J^+_f(A) & if \quad B \subseteq f(B) \\
J^+_f(A \cap B) & if \quad f(B) \subseteq B.
\end{array}
$$

Proof. By induction on n, we prove that $\forall n \geq 1, (B \to f)^n(A) = f^n(A) \cap \cap^n_{i=1} f^i(B)$.

– Basic case $n = 1$: by Def. 3.3, Prop. 3.30, and injectivity of f,

$$(B \to f)(A) = f(B \cap A) = f(A) \cap f(B).$$

– Inductive case: by injectivity of f,

$$
\begin{aligned}
(B \to f)^{n+1}(A) &= (B \to f)((B \to f)^n(A)) \\
&= (B \to f)(f^n(A) \cap \cap^n_{i=1} f^i(B)) \\
&= f^{n+1}(A) \cap \cap^{n+1}_{i=1} f^i(B).
\end{aligned}
$$

Thus, $J^+_{(B \to f)}(A) = \cup_{n \geq 1}(f^n(A) \cap \cap^n_{i=1} f^i(B))$. If $B \subseteq f(B)$, this expression becomes

$$\cup_{n \geq 1}(f^n(A) \cap f(B)) = f(B) \cap \cup_{n \geq 1} f^n(A) = f(B) \cap J^+_f(A).$$

If $f(B) \subseteq B$, we have by injectivity of f

$$\cup_{n \geq 1}(f^n(A) \cap f^n(B)) = \cup_{n \geq 1} f^n(A \cap B) = J^+_f(A \cap B).$$

Structure of Invariants.

Proposition 6.8. $J^{(B \to f)}$ *is full (resp. atomic) on a covering α iff J^f is full (resp. atomic) on the domain-restricted covering $(B \to \alpha)$.*

Proof. Using α and the domain-restricted covering $(B \to \alpha) = \{(B \to A_i)|A_i \in \alpha\}$, we have:

$$
\begin{aligned}
(B \to f)_i(A) & \\
&= ((B \to f) \cap A_i)(A) \\
&= (f \cap (B \times X) \cap A_i)(A) \\
&= (f \cap (B \to A_i))(A).
\end{aligned}
$$

6.3.2 Range Restriction

Kind. As in the previous case, range restrictions do not change anything to the "kind" of variant systems:

$$\kappa(f \leftarrow B) \in \{\kappa(f), 0\}.$$

Invariance.

Proposition 6.9. *If (X, f) is an injective RDS, then the greatest potential backward invariant of its range restriction $(f \leftarrow B)$ is*

$$J_-^{(f \leftarrow B)} \;=\; \begin{array}{ll} J_-^f(B) & \text{if } f(B) \subseteq B \subseteq f(X) \\ J_-^f & \text{if } f(B) \cup f(X) \subseteq B \\ B & \text{if } B \subseteq f(B). \end{array}$$

Proof. By induction on n, we prove that $\forall n \geq 1, (f \leftarrow B)^n(X) = f^n(X) \cap \cap_{i=0}^{n-1} f^i(B)$.

– Basic case $n = 1$: by Def. 3.5 and Prop. 3.30,

$$(f \leftarrow B)(X) = f(X) \cap B.$$

– Inductive case: by injectivity of f,

$$\begin{aligned} (f \leftarrow B)^{n+1}(X) &= (f \leftarrow B)((f \leftarrow B)^n(X)) \\ &= (f \leftarrow B)(f^n(X) \cap \cap_{i=0}^{n-1} f^i(B)) \\ &= f^{n+1}(X) \cap \cap_{i=0}^n f^i(B). \end{aligned}$$

Then, $(f \leftarrow B)^n(X)$ can be simplified into

$$\begin{array}{ll} f^{n-1}(B) & \text{if } f(B) \subseteq B \subseteq f(X) \\ f^n(X) & \text{if } f(B) \cup f(X) \subseteq B \\ B & \text{if } B \subseteq f(B). \end{array}$$

Proposition 6.10. *If (X, f) is an injective RDS, then the least potential forward invariant greater than A of its range restriction $(f \leftarrow B)$ is*

$$J_{(f \leftarrow B)}^+(A) = \cup_{n \geq 1}(f^n(A) \cap \cap_{i=0}^{n-1} f^i(B)).$$

Moreover,

$$J_{(f \leftarrow B)}^+(A) \;=\; \begin{array}{ll} B \cap J_f^+(A) & \text{if } B \subseteq f(B) \\ J_f^+(f(A) \cap B) & \text{if } f(B) \subseteq B. \end{array}$$

Proof. By induction on n, we prove that $\forall n \geq 1, (f \leftarrow B)^n(A) = f^n(A) \cap \cap_{i=0}^{n-1} f^i(B)$.

– Basic case $n = 1$: by Def. 3.5 and Prop. 3.30,

$$(f \leftarrow B)(A) = f(A) \cap B.$$

– Inductive case: by injectivity of f,

$$\begin{aligned} (f \leftarrow B)^{n+1}(A) &= (f \leftarrow B)((f \leftarrow B)^n(A)) \\ &= (f \leftarrow B)(f^n(A) \cap \cap_{i=0}^{n-1} f^i(B)) \\ &= f^{n+1}(A) \cap \cap_{i=0}^n f^i(B). \end{aligned}$$

Thus, $J^+_{(f \leftarrow B)}(A) = \cup_{n \geq 1}(f^n(A) \cap \cap_{i=0}^{n-1} f^i(B))$. If $B \subseteq f(B)$, this expression becomes

$$\cup_{n \geq 1}(f^n(A) \cap B) = B \cap \cup_{n \geq 1} f^n(A) = B \cap J^+_f(A).$$

If $f(B) \subseteq B$, by injectivity of f,

$$\cup_{n \geq 1}(f^n(A) \cap f^{n-1}(B)) = \cup_{n \geq 1} f^{n-1}(f(A) \cap B) = J^+_f(f(A) \cap B).$$

Structure of Invariants.

Proposition 6.11. $J^{(f \leftarrow B)}$ *is full (resp. atomic) on a covering α iff J^f is full (resp. atomic) on the range-restricted covering $(\alpha \leftarrow B)$.*

Proof. Using α and the range-restricted covering $(\alpha \leftarrow B) = \{(A_i \leftarrow B) | A_i \in \alpha\}$, we have:

$$(f \leftarrow B)_i(A)$$
$$= ((f \leftarrow B) \cap A_i)(A)$$
$$= (f \cap (X \times B) \cap A_i)(A)$$
$$= (f \cap (A_i \leftarrow B))(A).$$

6.4 Negation

The case of negation is problematic, which confirms the impression we have from Prop(s). 3.30 and 3.41. In Chap. 3, we proposed another unary operator called *external negation* (Def. 3.33). Let us examine it here as its anti-monotonicity yields interesting results.

Kind. It does not really make sense to compute the kind of the negation of a RDS since the result does not belong to this class anymore: if f is a closed subset of $X \times X$, its complement $\sim f$ is open unless specific topologies are used. The same argument applies to external negation $\neg f$.

Invariance. Generally, there is no short-cut to compute the invariant of a negated system. However, let us state a rather weak proposition about external negation.

Proposition 6.12. *Let (X, f) be a RDS. The greatest potential backward invariant of its external negation $\neg f$ is*

$$J^{\neg f}_- = X \backslash f(X).$$

Proof. Each $(\neg f)^i(X)$ with $i \geq 1$ is equal to an expression of the form

$$X \backslash f(X \backslash \cdots)$$

and we have:

$$X \setminus \cdots \subseteq X$$
$$\because \quad f \text{ monot.}$$
$$\Rightarrow \quad f(X \setminus \cdots) \subseteq f(X)$$
$$\because \quad \setminus \text{ anti-monot.}$$
$$\Rightarrow \quad X \setminus f(X \setminus \cdots) \supseteq X \setminus f(X)$$
$$\Rightarrow \quad \forall i, (\neg f)^i(X) \supseteq \neg f(X)$$
$$\Rightarrow \quad J_-^{\neg f} = X \setminus f(X).$$

Remark 6.13. It is interesting to observe the successive iterations of $\neg f$ from X. Below, each iteration is represented, and set inclusion appears from left to right, ending up with X:

$$\neg f(X) \subseteq (\neg f)^3(X) \subseteq \cdots \subseteq (\neg f)^{2n+1}(X) \subseteq \cdots$$
$$\cdots \subseteq (\neg f)^{2n}(X) \subseteq \cdots \subseteq (\neg f)^2(X) \subseteq X.$$

The odd powers of $\neg f$ are below the even powers. Thus, these successive iterations converge to the least fixpoint of $(\neg f)^2$ from below, and to its greatest fixpoint from above.

6.5 Sequential Composition

Kind. It is clear that the following proposition hold (we state it without proof).

Proposition 6.14. *Let (X, f) and (X, g) be two contracting (expanding) relations, then their sequential composition is also contracting (expanding):*

$$\kappa(f) = \kappa(g) \Rightarrow \kappa(f; g) = \kappa(f).$$

The contractivity factor of the composed systems is

$$\gamma(f; g) = \gamma(f) \cdot \gamma(g).$$

Invariance.

Proposition 6.15. *Let (X, f) and (X, g) be RDS. If f and g commute, i.e.*

$$\forall A \subseteq X, g(f(A)) = f(g(A)) \tag{Com}$$

then the greatest potential backward invariant of their sequential composition $f; g$ is

$$J_-^{f;g} = \begin{array}{ll} J_-^g(J_-^f) & \text{if} \quad g(J_-^f) \subseteq J_-^f \\ J_g^+(J_-^f) & \text{if} \quad J_-^f \subseteq g(J_-^f), \end{array}$$

and symmetrically for f and g swapped.

Proof. By induction on n, it is easy to prove that (Com) entails $(f;g)^n(X) = g^n(f^n(X)) = f^n(g^n(X))$. Thus, we have

$$J_-^{f;g} = \cap_i g^i(f^i(X)).$$

Since these iterations are decreasing, the invariant is computed by the limit of $g^i(f^i(X))$, which is equal to the limit of $g^i(J_-^f)$.

Proposition 6.16. *Let (X, f) and (X, g) be RDS. If (Com) holds, then the least potential forward invariant greater than A of their sequential composition $f;g$ is*

$$
\begin{array}{llll}
J_{f;g}^+(A) & = & J_g^+(J_f^+(A)) & \text{if} \quad (A \subseteq f(A)) \wedge (J_f^+(A) \subseteq g(J_f^+(A))) \\
& & J_-^g(J_f^+(A)) & \text{if} \quad (A \subseteq f(A)) \wedge (g(J_f^+(A)) \subseteq J_f^+(A)) \\
& & J_g^+(J_-^f(A)) & \text{if} \quad (f(A) \subseteq A) \wedge (J_-^f(A) \subseteq g(J_-^f(A))) \\
& & J_-^g(J_-^f(A)) & \text{if} \quad (f(A) \subseteq A) \wedge (g(J_-^f(A)) \subseteq J_-^f(A))
\end{array}
$$

and symmetrically for f and g swapped.

Proof. Again, assuming that (Com) holds, we have:

$$J_{f;g}^+(A) = \cup_i (f;g)^i(A) = \cup_i g^i(f^i(A))$$

from which we extract four cases appearing in the proposition.

Attraction. Here, we want to characterize the attraction of $f;g$ if we know something about the individual attractions: $P_1 \overset{f}{\rightsquigarrow} Q_1$ and $P_2 \overset{g}{\rightsquigarrow} Q_2$.

Proposition 6.17. *Let (X, f) and (X, g) be RDS. If (Com) holds, $P_1 \overset{f}{\rightsquigarrow} Q_1$, $P_2 \overset{g}{\rightsquigarrow} Q_2$, and $Q_1 \subseteq P_2$, then attraction is transitive, that is,*

$$P_1 \overset{f;g}{\rightsquigarrow} Q_2.$$

Proof. If f and g commute, i.e. if (Com) holds,

$$(f;g)^\omega(P_1) = g^\omega(f^\omega(P_1)) = g^\omega(Q_1).$$

Then, if $Q_1 \subseteq P_2$, the system clearly converges toward Q_2, or a part of it.

If a weaker property is verified, another proposition can be stated, the proof of which is left to the reader.

Proposition 6.18. *Let (X, f) and (X, g) be RDS. If (Com) holds and $Q_1 \cap J_+^g \cap P_2 \neq \emptyset$, then*

$$g^\omega(Q_1) \cap Q_2 \neq \emptyset.$$

6.6 Intersection

Kind. Concerning intersection, there is no general situation. All combinations can exist, and in the case of our relational dynamical systems, the result is often a single point, or a finite set of points, on which the contractivity factor does not really make sense anymore.

Invariance.

Proposition 6.19. *Let* (X, f) *and* (X, g) *be injective RDS, and*

$$\forall y \in Rg(f) \cap Rg(g), f^{-1}(y) \cap g^{-1}(y) \neq \emptyset. \tag{Int}$$

If (Com) holds, then the greatest potential backward invariant of their intersection $f \cap g$ *is*

$$
J_-^{f \cap g} = \begin{array}{ll}
J_-^f(J_-^g) & \text{if} \quad f(J_-^g) \subseteq J_-^g \\
J_-^g & \text{if} \quad J_-^g \subseteq f(J_-^g)
\end{array}
$$

and symmetrically when exchanging f *and* g. *If* $\forall A, f(A) \subseteq g(A)$ *holds, then* $J_-^{f \cap g} = J_-^f$.

Proof. Using a formal language notation to denote successive applications of f and g, we prove by induction on n that $\forall n, (f \cap g)^n(X) = \cap_{w \in \{f,g\}^n} w(X)$.

– Basic case $n = 1$: by (Int),

$$(f \cap g)(X) = f(X) \cap g(X).$$

– Inductive case: by (Int) and injectivity of f and g,

$$
\begin{aligned}
(f \cap g)^{n+1}(X) &= (f \cap g)(\cap_{w \in \{f,g\}^n} w(X)) \\
&= f(\cap_{w \in \{f,g\}^n} w(X)) \cap g(\cap_{w \in \{f,g\}^n} w(X)) \\
&= \cap_{w \in \{f,g\}^n} f(w(X)) \cap \cap_{w \in \{f,g\}^n} g(w(X)) \\
&= \cap_{w \in \{f,g\}^{n+1}} w(X).
\end{aligned}
$$

If (Com) holds, then $\cap_{w \in \{f,g\}^n} w(X) = \cap_{k=0}^n f^k(g^{n-k}(X))$. The particular cases are obtained easily.

If $\forall A, f(A) \subseteq g(A)$ holds, then $\forall w \in \{f,g\}^n, f^n(X) \subseteq w(X)$. Thus, $J_-^{f \cap g} = \cap f^n(X) = J_-^f$.

Proposition 6.20. *Let* (X, f) *and* (X, g) *be RDS. If (Com) and (Int) hold, and if* f *and* g *are injective, then the least potential forward invariant greater than* A *of their intersection* $f \cap g$ *is*

$$
J_{f \cap g}^+(A) = \begin{array}{ll}
J_-^f(J_-^g(A)) & \text{if} \quad (g(A) \subseteq A) \wedge (f(J_-^g(A)) \subseteq J_-^g(A)) \\
J_f^+(J_-^g(A)) & \text{if} \quad (g(A) \subseteq A) \wedge (J_-^g(A) \subseteq f(J_-^g(A))) \\
J_-^f(J_g^+(A)) & \text{if} \quad (A \subseteq g(A)) \wedge (f(J_g^+(A)) \subseteq J_g^+(A)) \\
J_f^+(J_g^+(A)) & \text{if} \quad (A \subseteq g(A)) \wedge (J_g^+(A) \subseteq f(J_g^+(A)))
\end{array}
$$

and symmetrically when exchanging f *and* g.

Proof. By induction on n, it is easy to prove that $\forall n, (f \cap g)^n(A) = \cap_{k=0}^n f^k(g^{n-k}(A))$. Thus, $J_{f \cap g}^+(A) = \cup_n \cap_{k=0}^n f^k(g^{n-k}(A))$. The four particular cases are summarized in the proposition.

Attraction.

Proposition 6.21. *Let (X, f) and (X, g) be RDS. If $P_1 \xrightarrow{f} Q_1$ and $P_2 \xrightarrow{g} Q_2$, then*

$$P_1 \cap P_2 \xrightarrow{f \cap g} Q_1 \cap Q_2.$$

Proof. The argument is based on monotonicity:

$$f^\omega(P_1) \subseteq Q_1$$
$$\Rightarrow \quad f^\omega(P_1 \cap P_2) \subseteq Q_1$$
$$\Rightarrow \quad (f \cap g)^\omega(P_1 \cap P_2) \subseteq Q_1$$
$$\because \quad \text{symmetrically with } Q_2$$
$$\Rightarrow \quad (f \cap g)^\omega(P_1 \cap P_2) \subseteq Q_1 \cap Q_2.$$

6.7 Union

Union of systems is a very important composition because many different types of combinations between systems and processes can be modeled this way.

Kind. It is not straightforward to compute the kind of union of systems. Nevertheless, some cases are very simple: the homogeneous ones

$$\kappa(f) = \kappa(g) \Rightarrow \kappa(f \cup g) = \kappa(f)$$

or the semi-neutral ones

$$\kappa(f) = 0 \Rightarrow \kappa(f \cup g) = \kappa(g).$$

The other cases cannot be solved automatically. A specific study is required to find the global kind.

Let us prove the first case, which is the most interesting [28].

Proposition 6.22. *Let (X, f) and (X, g) be two contracting RDS with contractivity factors $\gamma(f)$ and $\gamma(g)$. Their union is also contracting with a contractivity factor $\gamma(f \cup g) = \sup\{\gamma(f), \gamma(g)\}$.*

Proof. By Prop. 2.64, we know that $\forall A, B \subseteq X$,

$$h(f(A), f(B)) \leq \gamma(f) \cdot h(A, B)$$
$$h(g(A), g(B)) \leq \gamma(g) \cdot h(A, B).$$

We have, by Def. 3.15, Prop. 2.60 and the hypothesis, successively:

$$
\begin{aligned}
&h(f \cup g(A), f \cup g(B)) \\
= \ &h(f(A) \cup g(A), f(B) \cup g(B)) \\
\leq \ &\sup\{h(f(A), f(B)), h(g(A), g(B))\} \\
\leq \ &\sup\{\gamma(f), \gamma(g)\} \cdot h(A, B).
\end{aligned}
$$

Invariance. The assumptions needed to compute union invariants are not as strong as those required for intersection.

Proposition 6.23. *Let (X, f) and (X, g) be RDS. If (Com) holds, then the greatest potential backward invariant of their union $f \cup g$ is*

$$
J_-^{f \cup g} = \begin{array}{ll}
J_f^+(J_-^g) & \text{if} \quad J_-^g \subseteq f(J_-^g) \\
J_-^g & \text{if} \quad f(J_-^g) \subseteq J_-^g
\end{array}
$$

and symmetrically when exchanging f and g.

Proof. By induction on n, it can be proved that $\forall n, (f \cup g)^n(X) = \cup_{w \in \{f,g\}^n} w(X)$. By (Com), this can be rewritten $\forall n, (f \cup g)^n(X) = \cup_{k=0}^n f^k(g^{n-k}(X))$. Thus, $J_-^{f \cup g} = \cap_n \cup_{k=0}^n f^k(g^{n-k}(X))$. The particular cases are summarized in the proposition.

Proposition 6.24. *Let (X, f) and (X, g) be RDS. If (Com) holds, then the least potential forward invariant greater than A of their union $f \cup g$ is*

$$
J_{f \cup g}^+(A) = \begin{array}{ll}
J_f^+(J_g^+(A)) & \text{if} \quad (A \subseteq g(A)) \wedge (J_g^+(A) \subseteq f(J_g^+(A))) \\
J_g^+(A) & \text{if} \quad (A \subseteq g(A)) \wedge (f(J_g^+(A)) \subseteq J_g^+(A)) \\
J_f^+(J_-^g(A)) & \text{if} \quad (g(A) \subseteq A) \wedge (J_-^g(A) \subseteq f(J_-^g(A))) \\
J_-^g(A) & \text{if} \quad (g(A) \subseteq A) \wedge (f(J_-^g(A)) \subseteq J_-^g(A))
\end{array}
$$

and symmetrically when exchanging f and g.

Proof. It is similar to that of Prop. 6.23, because of the few assumptions required. So, we prove by induction

$$
J_{f \cup g}^+(A) = \cup_n \cup_{k=0}^n f^k(g^{n-k}(A)),
$$

and we get the four cases stated in the proposition.

Structure of Invariants. For the union of relations, many propositions can be proved, depending upon different combinations of properties. The most interesting results imply a drastic increase of complexity, by composition of elementary systems having fixpoint invariants.

The next theorems are all based on a finite union of relations $f = \cup_i f_i$. A unique symbol i is attached to each relation f_i. The alphabet Σ contains these symbols, and the covering is $\alpha = \{f_i | i \in \Sigma\}$.

Theorem 6.25 (Union invariant-1). *Let (X, f_i) be contracting RDS such that $\forall i, f_i^{-1}(X) = X$. Then, the greatest potential global invariant J of $(X, \cup_i f_i)$ is full and atomic. The function $\pi : \Sigma^{\mathbb{Z}} \mapsto J$ defined by $\pi(\sigma, \tau) = J_{\sigma,\tau}$ for all bi-infinite traces (σ, τ) is continuous.*

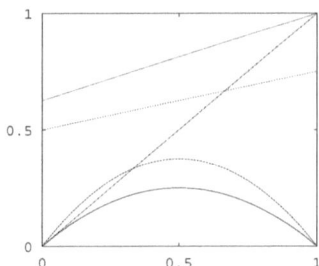

Fig. 6.1. Example of system verifying the assumptions of Theorem 6.25

Proof. Since each f_i is contracting, $\forall A \subseteq X$, we have

$$diam(f_i(A)) \leq \gamma(f_i) \cdot diam(A).$$

Thus, for any finite word $\sigma \in \Sigma^*$,

$$diam(f_\sigma(A)) \leq (\sup_i \gamma(f_i))^{|\sigma|} \cdot diam(A).$$

This right expression converges to 0 as $|\sigma|$ tends to infinity, because $\forall i, \gamma(f_i) < 1$. Since X is compact and complete, J_σ, $= f_\sigma(X)$ is not empty and it contains a unique element. By hypothesis, $\forall i, f_i^{-1}(X) = X$, which entails $J_{,\tau} = f_{,\tau}(X) = X, \forall \tau \in \Sigma^\omega$.

Each bi-infinite trace is associated to a unique element, which permits to define a function

$$\pi \ : \ \Sigma^{\mathbb{Z}} \mapsto J$$
$$\text{s.t.} \quad \pi(\sigma, \tau) = J_{\sigma,\tau}.$$

We have to prove that π is continuous: past traces only are taken into account, because of the last assumption again,

$$\forall \varepsilon, \forall \sigma, \exists \delta, \forall \sigma', d_a(\sigma, \sigma') < \delta \Rightarrow d(\pi(\sigma), \pi(\sigma')) < \varepsilon.$$

If $\sigma \neq \sigma'$, there exists n such that $\sigma|_{n-1} = \sigma'|_{n-1}$ but $\sigma_n \neq \sigma'_n$. The distance between these two traces is $d_a(\sigma, \sigma') = 2^{-n}$, and we rewrite the traces as follows: $\sigma = wu$ and $\sigma' = wv$. The invariant J_σ, is such that

$$J_\sigma, = f_\sigma, = f_w,(f_u,(X)) \subseteq f_w,(X);$$

the same argument applies for $J_{\sigma'}, \subseteq f_{w_{,}}(X)$. Hence, we have

$$d(J_{\sigma,}, J_{\sigma',}) \leq diam(f_{w_{,}}(X)) \leq (\sup_i \gamma(f_i))^{|w|} \cdot diam(X)$$

where $|w| = n - 1$. Thus, if we know ε, it suffices to compute $n = 1 + \frac{\ln(\varepsilon) - \ln(diam(X))}{\ln(\sup_i \gamma(f_i))}$ and δ is simply 2^{-n} to get continuity.

Theorem 6.26 (Union invariant-2). *Under the assumptions of Theorem 6.25, if in addition $\sum_i \gamma_i(f_i) < 1$, then the greatest potential global invariant J of $(X, \cup_i f_i)$ is totally disconnected.*

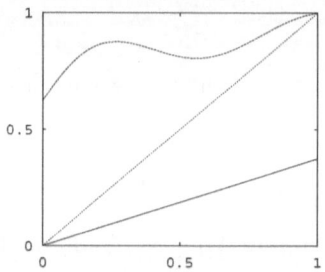

Fig. 6.2. Example of system verifying the assumptions of Theorem 6.26

Proof. Let x and y be in J. We choose p such that

$$diam(J) \cdot (\sum_i \gamma(f_i))^p < d(x, y).$$

Since $J = J_- \cap J_+$ is invariant, and $J_+ = X$ by the assumption $\forall i, f_i^{-1}(X) = X$, we have $J \subseteq \cup_{w \in \Sigma^p} f_w(J)$ by induction on p.

– Basic case $p = 1$: by definition of J_- and $J = J_-$,

$$J \subseteq f(J) = \cup_i f_i(J).$$

– Inductive case: on case p, we apply the basic case; by monotonicity and universal disjunctivity of all f_i's, this gives

$$
\begin{aligned}
J &\subseteq \cup_{w \in \Sigma^p} f_w(J) \\
&\subseteq \cup_{w \in \Sigma^p} f_w(\cup_i f_i(J)) \\
&= \cup_{w \in \Sigma^p} \cup_i f_{iw}(J) \\
&= \cup_{w \in \Sigma^{p+1}} f_w(J).
\end{aligned}
$$

We know that for any word in Σ^p,

$$diam(f_w(J)) \leq (\prod_i \gamma(f_{w_i})) \cdot diam(J).$$

Let us consider all such components of the invariant:

$$\sum_{w \in \Sigma^p} diam(f_w(J))$$

$$\leq \sum_{w \in \Sigma^p} (\prod_i \gamma(f_{w_i})) \cdot diam(J)$$

$$= (\sum_i \gamma(f_i))^p \cdot diam(J)$$

$$< d(x, y).$$

States x and y are thus further apart from each other than the diameters of all components of the invariant. Hence, they must be in distinct such components, that can be as small as desired. Thus, the invariant set is totally disconnected.

Theorem 6.27 (Union invariant-3). *Under the assumptions of Theorem 6.25, if the individual relations f_i are injective, and if there exist $i \neq j$ such that the invariants of f_i and f_j are two different fixpoints, then the greatest potential global invariant J of $(X, \cup_i f_i)$ is perfect.*

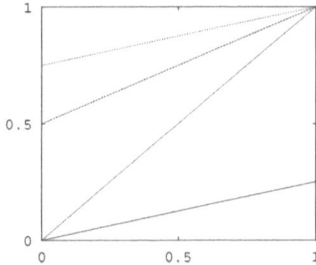

Fig. 6.3. Example of system verifying the assumptions of Theorem 6.27

Proof. Suppose J is not perfect: $\exists \sigma, \tau \in \Sigma^\omega$ such that $\pi(\sigma, \tau)$ is isolated in J. Since π is continuous, there must be a δ such that $\pi(\sigma', \tau) = \pi(\sigma, \tau)$ for all σ' such that $d_a(\sigma, \sigma') < \delta$. Thus, for n sufficiently large, we set $u = \sigma|_n iii \cdots$ and $v = \sigma|_n jjj \cdots$, so that $\pi(u, \tau) = \pi(v, \tau) = \pi(\sigma, \tau)$. The first equality implies

$$f_{u_\cdot}(X) = f_{v_\cdot}(X)$$
$$f_{\sigma|_n \cdot}(J^{f_i}) = f_{\sigma|_n \cdot}(J^{f_j})$$

and, by injectivity of all f_i,

$$J^{f_i} = J^{f_j}$$

which contradicts the assumption.

Theorem 6.28 (Union invariant-4). *Under the assumptions of Theorem 6.25, if there exists $A \subseteq X$ such that $f(A) \subseteq A$, all f_i are injective on A, and $\forall i \neq j, f_i(A) \cap f_j(A) = \emptyset$, then the greatest potential global invariant J of $(X, \cup_i f_i)$ is totally disconnected and perfect, and the function π is injective.*

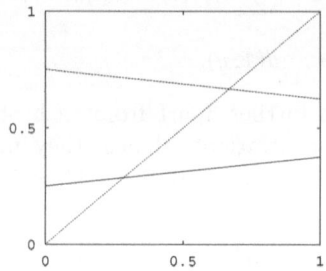

Fig. 6.4. Example of system verifying the assumptions of Theorem 6.28

Proof. Since $f(A) \subseteq A$, this last set contains the invariant $J \subseteq A$. The disjointness assumption applies to J. Injectivity of the f_i on A applies to J, too. Thus, the function $\pi : \Sigma^\omega \times \Sigma^\omega \mapsto J$ is injective. Indeed, since the forward images of X are mutually disjoint, this disjointness property remains valid for successive iterations:

$$\forall u \neq v \in \Sigma^n, f_u(X) \cap f_v(X) = \emptyset.$$

(This statement is easily proved by induction on n.) Hence, for u and v two distinct infinite words, there exists n such that $u|_n \neq v|_n$, and $f_{u|_n}(X) \cap f_{v|_n}(X) = \emptyset$. Since $J_u, \subseteq f_{u|_n}(X)$ and $J_v, \subseteq f_{v|_n}(X)$, these trace-parametrized invariants must be different, which implies injectivity of π.

The set of traces is totally disconnected and perfect, and its image by an injective function has the same properties.

Corollary 6.29 (Union invariant-5). *Under the assumptions of Theorem 6.28, or combining the assumptions of Theorems 6.26 and 6.27, the greatest potential global invariant J of $(X, \cup_i f_i)$ is a Cantor set.*

Proof. The invariant is closed (Prop. 5.8), totally disconnected and perfect (Theorem 6.28). It is thus a Cantor set (Def. 2.28).

Remark 6.30. – The previous results (6.25 to 6.29) have been obtained essentially by a conjunction of different results from fractal theory: [328, Theorems A, D], [159, §§3.1(3), (9)], [140, Theorem 4.3], and [325, Prop. 3.3.4, Cor. 3.3.5]. In the last reference, the author rederives many results using nonstandard analysis.
 – Under the global contraction assumptions of Theorems 6.25-6.29, convergence could also be reached by relaxing the dynamics to fair, i.e. pseudo-periodic asynchronous, iterations (see pp. 50 and 78).

The following proposition gives under- and upper- approximations of the invariant Cantor set produced by an iterated union.

Proposition 6.31. *Under the assumptions of Theorem 6.28 with two component relations f and g, the global invariant of $f \cup g$ lies in a subspace of X determined by the individual invariants and their images: this subspace is bounded by $\min(J^f, J^g, f(J^g), g(J^f))$ and $\max(J^f, J^g, f(J^g), g(J^f))$.*

Proof. Relation f alone attracts everything to J^f from everywhere in X, even from J^g. applying f twice from J^g will be closer from J^f than applying it once. Thus, to get the limits of the Cantor set, it suffices to consider the individual invariants and their images by the other relation.

Attraction. Here, it is interesting to give some counterexamples to very simple statements, showing how complex union can be, thereby confirming our previous results on this operator.

Let us imagine that $P_1 \overset{f}{\rightsquigarrow} Q_1$ and $P_2 \overset{g}{\rightsquigarrow} Q_2$. We would like to have a proposition like Prop. 6.21, i.e.

$$P_1 \cup P_2 \overset{f \cup g}{\rightsquigarrow} Q_1 \cup Q_2.$$

Nothing can be said because we do not know how f behaves from P_2, nor g from P_1.

Adding the assumption $P_1 = P_2$ does not change the result since we do not know how f behaves from $g(P_1)$.

Finally, adding the assumption $Q_1 = Q_2$, we get the announced result but it is not very useful!

Actually, as shown by Theorems 6.25 to 6.36, we need to add some assumptions in order to get something interesting out of union, because the behavior of this operator can lead to very rich structures. For instance, these theorems can be restated to emphasize attraction by invariant. In this case, we know that the attractor has a Cantor set structure, and we know its maximal boundaries.

6.8 Products

6.8.1 Free Product

For the free product, the situation is much simpler than for the previous composition operators, since there is no interaction between the components composed together.

Kind. By definition, the kind of a free product of systems is obtained by juxtaposition of the individual kinds in a multidimensional vector. For instance, in the binary case:

$$\kappa(f \times g) = (\kappa(f), \kappa(g)).$$

This allows to separate the reasoning on each component whenever possible.

Invariance. No assumptions are needed to compute the invariants of a free product, due to the clear separation of components.

Proposition 6.32. *The invariant of a product of systems is equal to the product of the invariants of its components:*

$$
\begin{aligned}
J_-^{f \times g} &= J_-^f \times J_-^g \\
J_{f \times g}^+(A \times B) &= J_f^+(A) \times J_g^+(B).
\end{aligned}
$$

Proof. The two equalities are based on the following: $\forall n, (f \times g)^n(A \times B) = f^n(A) \times g^n(B)$. We prove it by induction on n.

− Basic case $n = 0$:

$$(f \times g)^0(A \times B) = A \times B.$$

− Inductive case:

$$
\begin{aligned}
(f \times g)^{n+1}(A \times B) &= (f \times g)(f^n(A) \times g^n(B)) \\
&= f^{n+1}(A) \times g^{n+1}(B).
\end{aligned}
$$

Structure of Invariants. The covering used to observe the dynamics of a system defined on a Cartesian product of spaces has to be decomposable to allow a decomposition of the proof of properties: if h is defined on $X \times Y$, the covering α defined on it has to be the product of two coverings, α_X on X and α_Y on Y, respectively associated to two alphabets Σ_X and Σ_Y. Thus, to each symbol i associated to a part A_i of α, a pair of symbols $(j, k) \in \Sigma_X \times \Sigma_Y$ must be associated such that j defines a part A'_j of α_X, k defines a part A''_k of α_Y, and $A_i = A'_j \times A''_k$. Let us denote this bijective correspondence as follows: $i \div (j, k)$. Hence, we have the following proposition.

Proposition 6.33. $J^{f \times g}$ *is full (atomic) on* $\alpha = \alpha_X \times \alpha_Y$ *if* J^f *is full (atomic) on* α_X *and* J^g *is full (atomic) on* α_Y.

Proof. We compute one iteration step, which is sufficient to reach the announced result:

$$(f \times g)_i(A \times B)$$
$$= ((f \times g) \cap A_i)(A \times B)$$
$$= ((f \times g) \cap (A'_j \times A''_k))(A \times B)$$
$$= ((f \cap A'_j) \times (g \cap A''_k))(A \times B)$$
$$= (f \cap A'_j)(A) \times (g \cap A''_k)(B)$$
$$= f_j(A) \times g_k(B).$$

Attraction. Again, the free product offers a very simple way to prove attraction: it suffices to prove the same property for each component independently. The following proposition states this in a clear way, the proof of which is left to the reader.

Proposition 6.34. *If* $P_1 \xrightarrow{f} Q_1$ *and* $P_2 \xrightarrow{g} Q_2$, *then*

$$P_1 \times P_2 \xrightarrow{f \times g} Q_1 \times Q_2.$$

6.8.2 Connected Product

Regarding the properties related to the general connected product $\otimes_R g_i$, things are much more difficult than for the free product. The problem clearly comes from the possible interaction between components, as described by R and by the g_i's:

$$\otimes_R g_i(\times_i A_i) \neq \times_i g_i(A_i).$$

In fact, no short-cut is possible in general, and a case by case study as to be carried out. To illustrate this, let us consider a simple example.

Example 6.35. Take two relations $f \in \mathcal{R}(X, V)$ and $g \in \mathcal{R}(Y, W)$. Their connected product $f \otimes_R g$ can be represented by $R(1) = \{1\}$ and $R(2) = \{1, 2\}$.

Here follow the first iteration steps leading to the greatest potential backward invariant:

$$\begin{aligned}
1: \quad & (f \otimes_R g)(V \times W) \\
& = f(\Pi_{R(1)}(V \times W)) \times g(\Pi_{R(2)}(V \times W)) \\
& = f(V) \times g(V \times W) \\
2: \quad & (f \otimes_R g)^2(V \times W) \\
& = f^2(V) \times g(f(V) \times g(V \times W)).
\end{aligned}$$

Certain classes of connected products could be analyzed in a compositional way, depending upon whether this sort of expression could be easily simplified or not. This is left for future work.

6.9 Combining Union with Free Product

This section briefly updates the list of union invariant theorems presented in §6.7, by considering unions of free products of systems.

Theorem 6.36 (Union invariant-6). *Let (X, f) and (X, g) be two compatible RDS, viz. $\kappa(f) = \kappa(g)$, with distinct full and atomic greatest potential global invariants. Assume $f = \times_i f_i$, $g = \times_i g_i$, and $\forall i$,*

- *(X_i, f_i) and (X_i, g_i) are injective RDS,*
- *either $\gamma(f_i) + \gamma(g_i) < 1$ and $f^{-1}(X_i) = g^{-1}(X_i) = X_i$,*
  or $\gamma(f_i^{-1}) + \gamma(g_i^{-1}) < 1$ and $f(X_i) = g(X_i) = X_i$.

Then $f \cup g$ has a full and atomic greatest potential global invariant with a Cantor-set structure.

Proof. Since f and g can be decomposed into relations that pairwise verify the assumptions of Theorems 6.25, 6.26, and 6.27, the result is obtained using Prop(s). 6.32 and 6.33.

Let us rephrase this theorem informally. Complexity (fullness and atomicity) is preserved by union composition of compatible systems having different complex invariants. An interesting particular case of this theorem can be stated when the individual components are degenerate complex systems, i.e. the individual invariants are just fixpoints as in Cor. 6.29. In this case, the result of union composition is also a complex system having a Cantor-set structure.

Corollary 6.37 (Union invariant-7). *Let (X, f) and (X, g) be two compatible RDS, that is, $\kappa(f) = \kappa(g)$, with distinct greatest potential global invariants, such that $f = \times_i f_i$, $g = \times_i g_i$, and $\forall i$,*

- *(X_i, f_i) and (X_i, g_i) are injective RDS,*
- *$\gamma(f_i) + \gamma(g_i) < 1$ and $f^{-1}(X_i) = g^{-1}(X_i) = X_i$*
  or $\gamma(f_i^{-1}) + \gamma(g_i^{-1}) < 1$ and $f(X_i) = g(X_i) = X_i$,
- *J^{f_i} and J^{g_i} are fixpoints invariants,*

then $f \cup g$ has a full and atomic greatest potential global invariant with a Cantor-set structure.

6.10 Discussion

In this section, we first summarize the compositional results of this chapter, and discuss the main assumptions needed to achieve them; second, the limitations and open problems of the approach are emphasized; third, we present some related work; finally, we recall informally how complexity can emergence by composition.

6.10.1 Compositionality: Summary

After the elaboration of a framework including relational dynamical systems, composition operators leading to structured systems, and the presentation of their dynamical properties, we investigated here the main question of this monograph, that is, the compositional analysis of dynamical systems, or in other words, the link between the structure of systems and their dynamical properties.

This led to original results allowing to deduce systematically global dynamical properties (e.g. invariants, and their structure) from local ones. They are summarized in Table 6.1: "ok" means that compositionality holds; "/" means that we could not derive interesting results by compositional analysis assuming reasonable hypotheses; otherwise, the main assumption(s) needed to achieve compositionality are recalled.

Table 6.1. Compositional analysis of dynamical properties: summary

Operator	Invariance	Structure	Attraction
$^{-1}$	ok	α^{-1}	repulsion
$(\rightarrow)$	injectivity	$(\rightarrow \alpha)$	/
$(\leftarrow)$	injectivity	$(\alpha \leftarrow)$	/
$\sim, \setminus$	/	/	/
$\neg$	ok	/	/
;	(Com)	/	(Com)
$\cap$	(Int) $\wedge$ (Com)	/	ok
$\cup$	(Com)	ok	/
$\times$	ok	$\alpha_X \times \alpha_Y$	ok
$\otimes$	/	/	/

In particular, we carried out the in-depth analysis of union. The diagram of Fig. 6.5 summarizes the results and shows their interrelations: each arrow stands for a logical implication, due to specific assumptions. As we shall see in Chap. 7, union is indeed fundamental in the study of many classical examples, like Smale's horseshoe map, Cantor's middle-thirds relation, and the logistic map.

Finally, we have evidenced a generic way to get complexity out of simple systems (see Cor. 6.37): it suffices to compose these systems in an adequate way, under some conditions: provided the subsystems have compatible dynamics attracting the global state space to different structured invariants, complexity emerges from their composition.

6.10.2 Limitations and Open Problems

Table 6.1 is useful to emphasize some current limitations and open problems of the compositional analysis of dynamical properties we have developed.

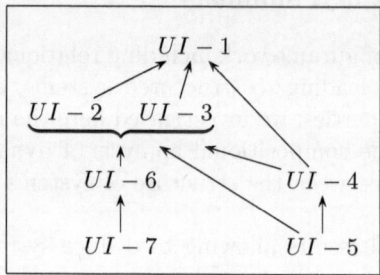

Fig. 6.5. Union-invariant theorems: summary

Negation and Difference. These operators are quite difficult to study in a compositional way, the main reason being they do not preserve the fundamental characteristic of RDS (closed relations), apart from specific topologies where sets are both open and closed. For the same reason, this limitation is not too restrictive. Nevertheless, the special anti-monotonic case of external negation should be analyzed in greater details.

Connected Product. This operator is also difficult to analyze in general, although it is essential in the construction of many interesting families of dynamical systems: neural networks [116, 290], cellular automata (see Chap. 8), etc.

The principal source of problems in compositional analysis of connected products is the explicit interaction between components described by the neighborhood relation. This relation can take many aspects and its structure strongly influences the global behavior of the product.

To overcome this limitation, we study particular cases of connected products, which offer promising results.

- In Chap. 8, attraction properties and specific compositions of cellular automata are studied. There, a homogeneous neighborhood relation limits the interaction to a small number of neighbors.
- In Chap. 9, computational properties of a general model based on the connected product are analyzed by composition. This model can be instanciated to Turing machines, cellular automata and continuous functions.

Structure of Invariants. In some cases, e.g. for sequential and intersection compositions, the structure of invariants cannot be obtained by simple composition of the invariant structure of their components.

Weaker versions of fullness and atomicity should be proposed and investigated, provided they still lead to qualitatively equivalent properties.

For example, instead of requiring the relation between traces and states to be universally satisfied, i.e. $\forall (\sigma, \tau) \in \Sigma^{\mathbb{Z}}, \# J_{\sigma, \tau} \cdots$, we could ask for a sufficient density of such traces, e.g. $\forall (\sigma, \tau) \in A, \# J_{\sigma, \tau} \cdots$, with A being dense in $\Sigma^{\mathbb{Z}}$. The challenge would then consist in preserving this density among composition.

Attraction. This property is often difficult to analyze, in a great majority of useful composition operators. This type of limitation is also mentioned in parallelism semantics, where the authors acknowledge the difficulty to achieve liveness in a compositional way, as far as interesting composition operators are concerned, i.e. where components can interact (union, synchronization, connected product) [1, 67].

Thus, the main challenge for the future of compositional analysis of dynamical systems is probably a deeper study of attraction properties, including reachability and potential attraction.

6.10.3 Related Work

Studying systems by decomposing them into simpler pieces and composing the individual results so as to obtain the global analysis is not new. Some ideas have already appeared in different disciplines, not always explicitly mentioning composition as a method.

General Scheme. A general hierarchical decomposition of complex systems is presented in [69].

Rewriting Systems. For some automata and term rewriting systems, termination, viz. finite-time attraction, is analyzed by composition in [94, 86, 247].

Theory of Programs. Transition systems analysis, parallel programs design and construction can be helped using a compositional approach [329, 19, 179, 299, 1, 67]. A relational framework for studying the semantics of programs is based on composition operators in [245, 246].

Dynamical Systems. Dynamical systems have not been studied by a compositional approach very often. Only a few references can give some first steps, e.g. [265, 300]. Nevertheless, in the context of fractals, composition appears implicitly in [328, 159, 140, 28, 325].

Cellular Automata. Cellular automata are sometimes regarded under algebraic means as composed systems, see [113, 317] for a detailed study of these aspects. Moreover, compositional aspects closer to our ideas have been proposed in [53, 316, 49].

Neural Networks. These systems can be seen as instances of connected products. Only a few authors have tried to study complex neural nets as composed from elementary systems. The principal references here are [251, 252, 253, 254, 255, 318, 319, 321].

Algorithmic Information Theory. This theory defines the complexity of a finite symbol sequence as the size of the smallest program able to generate this sequence [43, 56, 84, 199].

In [85] the author studies the complexity of sequences under composition. The results stated apparently contradict our results in that they generally express a decrease of complexity when sequence are composed together. This is normal because a common pattern or part of two component sequences can be generated using the same subprogram, which reduces the global complexity.

To reach the same results using our approach, it would be very interesting to evaluate the complexity of observed sequences as the compositional or structural complexity of generative systems, instead of looking at properties of invariant and attractors.

Preliminary Steps of the Approach. In [286], the first steps toward the present technical framework are given, including a discussion of potential developments including compositional properties. These ideas have then been developed in [116, 289, 101, 290].

6.10.4 Emergence of Complexity by Structural Composition

In this chapter, we examined several composition operators applied to dynamical properties such as invariance, structure of invariants, and attraction.

Let us summarize here (see Table 6.2) their behavior w.r.t. complexity, that is, properties of invariants and attractors. In general, complexity can be measured by the diversity of traces a system can show. Fullness and atomicity, as well as the number and the structure of attractors, are important indices of this kind of behavior.

Table 6.2. Complexity w.r.t. composition operators

Operators	Complexity
$^{-1}, (\rightarrow), (\leftarrow), ;, \cap$	$=$ or $\searrow$
$\cup, \times, \otimes$	$=$ or $\nearrow$
$\sim, \backslash$	?

- Some operators cannot increase complexity: inversion, restrictions, sequential composition, and intersection.
- Some operators cannot decrease complexity: union, free product.
- Negation can lead to a totally different system; it can thus dramatically increase complexity, but it can also decrease complexity.
- As to connected product, no general proof could be given but we suspect that, assuming non-trivial components and neighborhood relations, the operator falls in the second group: they cannot decrease complexity.

Let us come back to the second group and concentrate on union and free product. These operators are very simple: union corresponds to set-theoretic union of images of some systems, free product is a Cartesian product of relations, that is, without any interaction between components. However, despite their simplicity, they can generate complex behaviors, and we will see in the next chapter that they prove most useful in the analysis of some classical examples of complex systems.

The general way to generate complexity using these operators is to compose systems having compatible behaviors on the same subspaces (viz. axes), but attracting (in the future or in the past) the global state space to different (full and atomic) invariants.

In the following chapter (Chap. 7), we will show that some formal systems also fulfill the conditions for emergence of complexity, without any modification of the theory presented above.

Cellular automata will then be studied (see Chap. 8), as examples of connected products, but to show that complexity can arise for the same reasons as the ones stated above, we will have to add some physical measures to our theorems. Nevertheless, using these additional complexity measures, we will arrive at the same conclusions.

7. Case Studies: Compositional Analysis of Dynamics

This chapter is devoted to four case studies in the compositional analysis of dynamical properties. The first three studies are classical prototypes of complex systems: Smale horseshoe map, Cantor relation, logistic map. The last one is a well-known formal system generating paperfolding sequences.

The analysis of complex dynamical systems can be carried out by composition in a clear and effective way, whereas, classically, the analysis of these systems can take several pages of cumbersome developments.

In the examples, the compositional analysis yields known results. We show how to combine different types of composition described up to now, and their properties. We also see that a top-down analysis of dynamical systems is possible with an appropriate decomposition of the system into simpler ones.

The four systems are regarded as union and product compositions of elementary systems. In each case, using the results of Chap. 6, the complex dynamics can be explained by a rich structure of the respective invariants, which are full and atomic.

The present chapter is organized as follows: we briefly introduce our examples in §7.1; we successively analyze the Smale horseshoe map, the Cantor relation, and the logistic map in §§7.2–7.4; paperfoldings are then studied in §7.5; finally, we close the chapter with a discussion in §7.6.

7.1 A Collection of Complex Behaviors

Let us briefly explain the reasons why the examples we have chosen deserve attention. By the way, we also refer the interested reader to [138, 118, 279] for the missing historical references that complete our presentation.

Smale Horseshoe Map. This two-dimensional system is important because it constitutes a paradigmatic example of chaotic behavior. Indeed, sensitivity to initial conditions results from successive stretching and folding of the state space. This process yields mixing via filamentation, as a baker who kneads a blob dough [293]. Many other chaotic systems like Lorentz equations [207] or Hénon map [145], are also subject to mixing via such spatial deformations, which seem to be important factors of complexity (see also Ex(s). 3.23 and 5.65, which show the same behavior). We analyze this phenomenon by

F. Geurts: Abstract Compositional Analysis of Iterated Relations, LNCS 1426, pp. 163-181, 1998.
© Springer-Verlag Berlin Heidelberg 1998

rewriting the system as a union of two products, and explain it by using the union-invariant theorems detailed in Chap. 6.

Cantor Relation. This system is interesting because it generates the well-known Cantor middle-thirds set as invariant. This set is a typical example of fractal, and its properties have been subject to many studies for more than a hundred years [96, 28]. Despite its nondeterminism, it can be treated easily in our relational framework: it suffices to reverse the execution, and to consider past invariance instead of future invariance. Of course, the global invariant takes both directions into account, which permits to hide any consideration of time direction. Union is again the appropriate composition operator to study this system.

Logistic Map. This is a typical example of one-dimensional chaotic system: every textbook on dynamical systems uses it as running example [28, 88, 154]. Again, we analyze it compositionally. However, our objective here slightly differs from the first examples: we merely want to show that the compositional analysis of the logistic map can be obtained as a by-product of the analysis of the Cantor relation. We voluntarily remain at an informal level, but we give intuitively clear transformation steps allowing to transfer the analysis from one system to the other one.

Paperfoldings. Finally, we concentrate on the dynamics of formal systems. A specific family serves as introductory example: paperfoldings. These seemingly funny systems are important for three reasons.

The fractal curves they generate can be used in fractal image compression processes, and correspond to natural morphogeographic patterns [279].

The folding process physically realizes an equivalent transformation to Smale's "stretch-and-squeeze" horseshoe map. It is thus not surprising to find the same complexity in paperfolding sequences as in the behavior of horseshoe-like maps.

Studying formal systems as dynamical systems is not frequent [73, 77, 268], even though formal systems are often used to characterize dynamical properties of systems (e.g. see §5.7.1 and [205, 11]). This simple but promising example illustrates how apparently disjoint fields can enrich each other. Many formal systems could be analyzed by composition of dynamical systems: formal grammars, L-systems, dynamic proofs, etc.

7.2 Smale Horseshoe Map

The first example we analyze is a simplified version of the classical "Smale horseshoe map" [292, 293]. We suggest the interested reader to follow [326, §4.1, pp. 420–437] in parallel to compare a classical analysis with our compositional analysis.

The System. Before presenting the system itself, let us fix some parameters. Let $0 < \lambda < \frac{1}{2}$ and $0 < \frac{1}{\mu} < \frac{1}{2}$. We choose $\lambda = \frac{1}{3}$ and $\frac{1}{\mu} = \frac{1}{3}$.

The function f, defined on $[0, 1]^2$, reads

$$f(x, y) = \begin{cases} (\lambda x, \mu y) & \text{on} \quad H_0 = A_x \times B_y \\ (-\lambda x + 1, -\mu y + \mu) & \text{on} \quad H_1 = A_x \times C_y \end{cases}$$

with the following notational conventions:

$$
\begin{aligned}
A_x &= [0, 1] \\
A_y &= [0, 1] \\
B_y &= [0, \frac{1}{\mu}] \\
C_y &= [1 - \frac{1}{\mu}, 1] \\
D &= A_x \times A_y;
\end{aligned}
$$

moreover, two horizontal rectangles are defined in D:

$$
\begin{aligned}
H_0 &= A_x \times B_y \\
H_1 &= A_x \times C_y.
\end{aligned}
$$

The domain $D = [0, 1] \times [0, 1]$ is represented in Fig. 7.1.

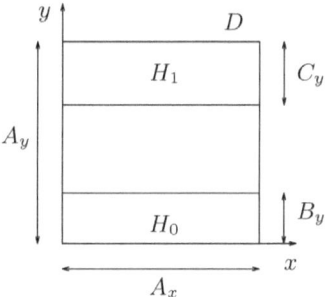

Fig. 7.1. Domain decomposition

The graphical representation of f appears in Fig. 7.2. The shape of the codomain explains why f is said to be horseshoe-like. A second iteration from D is also represented in Fig. 7.3. Observe the ongoing filamentation.

An Adequate Decomposition. We remark that the state spaces are disjoint: x and y act independently. How can we decompose f into simpler components? An easy way to do it is the following:

$$f = f_0 \cup f_1$$

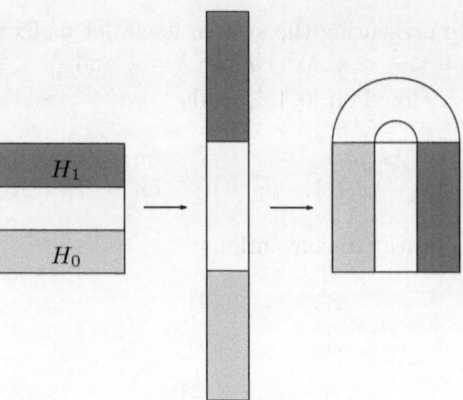

Fig. 7.2. Effect of f on $[0,1]^2$: y-stretch, x-squeeze, fold

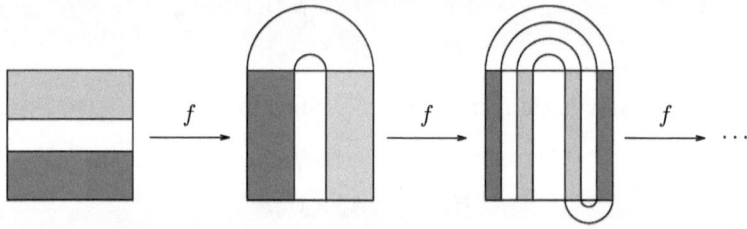

Fig. 7.3. Successive iterations of f from $[0,1]^2$

with

$$
\begin{aligned}
f_0 &= R \times S && \text{on} && H_0 \\
f_1 &= V \times W && \text{on} && H_1
\end{aligned}
$$

and (see also Fig. 7.4):

$$
\begin{aligned}
R(x) &= \lambda x && \text{on} && A_x \\
S(y) &= \mu y && \text{on} && B_y \\
V(x) &= -\lambda x + 1 && \text{on} && A_x \\
W(y) &= -\mu y + \mu && \text{on} && C_y.
\end{aligned}
$$

Individual Analyses. Since $\lambda < \frac{1}{2}$, components R and V are contracting in the future. Their contractivity factors are

$$
\gamma(R) = \gamma(V) = \lambda.
$$

In the same way, since $\mu > 2$, S and W are expanding in the future, with contractivity factors

$$
\gamma(S) = \gamma(W) = \mu.
$$

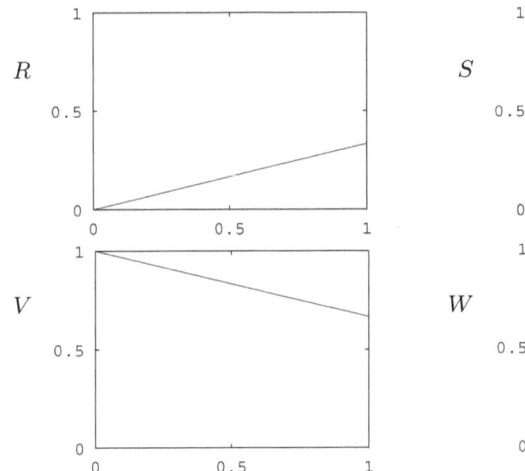
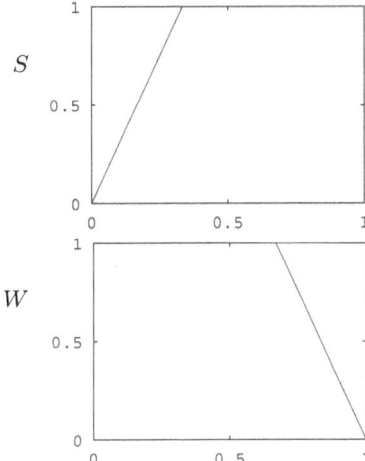

Fig. 7.4. Graphs of R, S, V, and W

Thus, the individual kinds are

$$\kappa(R) = \kappa(V) = -$$

and

$$\kappa(S) = \kappa(W) = +.$$

The individual invariants are easily computed. They are simple fixpoints:

$$
\begin{aligned}
J^R &= 0 \\
J^S &= 0 \\
J^V &= \frac{1}{1+\lambda} \\
J^W &= \frac{\mu}{1+\mu}.
\end{aligned}
$$

Compositional Analysis. The two products $R \times S$ and $V \times W$ are *hyperbolic systems*, i.e. systems which are contracting in opposite temporal directions (past or future) on different axes.

However, the two subsystems f_0 and f_1 have compatible dynamics,

$$\kappa(f_0) = \kappa(f_1) = (-, +).$$

Their invariants are distinct fixpoints:

$$
\begin{aligned}
J^{f_0} &= (0,0) \\
J^{f_1} &= \left(\frac{1}{1+\lambda}, \frac{\mu}{1+\mu}\right),
\end{aligned}
$$

by Prop(s). 6.32 and 6.33.

To characterize the global invariant J^f of f, and its structure, we can thus use Cor. 6.37 whose assumptions are verified. The conclusion is: the invariant of the Smale horseshoe map f is full, atomic, and has a Cantor set structure.

Partial Conclusion. Let us summarize the analysis of this first example. We have here a structurally simple system principally based on a combination of linear pieces, union and free product, which shows a complex dynamics. Our approach offers a short and clear analysis, though many complicated proofs appear in the literature. Actually, it is "too simple" for the approach because it is only functional, there is no real relation involved. Let us turn to the next example, based on a relation.

7.3 Cantor Relation

The next example is a relation generating the Cantor middle-thirds set as attracting invariant set. We have already presented this set (see Ex. 2.29) and the way to obtain it (Ex(s). 5.49, 5.62, 5.64). In these examples, we have proved that the relation generating this set has a full, atomic, invariant set, using criteria defined in Chap. 5. Here, we will obtain the same result by compositional analysis of the system.

The System. System f is built on two elementary functions defined on the interval $[0, 1]$, represented in Fig. 7.5 and formally defined as follows:

$$f_0(x) \quad = \quad \frac{x}{3}$$
$$f_1(x) \quad = \quad \frac{x}{3} + \frac{2}{3}.$$

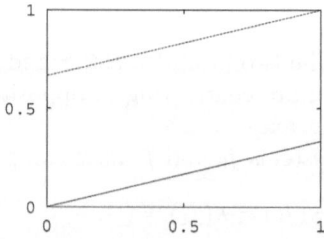

Fig. 7.5. Graph of $f_0 \cup f_1$

The union of these two functions leads to the Cantor relation:

$$f = f_0 \cup f_1.$$

Individual Analyses. The kinds of f_0 and f_1 are identical: they are both contracting in the future, i.e.

$$\kappa(f_0) = \kappa(f_1) = -.$$

The contractivity factors are:

$$\gamma(f_0) = \gamma(f_1) = \frac{1}{3}.$$

Function f_0 has a fixpoint invariant, $J^{f_0} = 0$, and function f_1 has also a fixpoint invariant, $J^{f_1} = 1$. These invariants are different.

Compositional Analysis. All assumptions of Theorem 6.28 and its corollary (Cor. 6.29) are verified. Hence we get as invariant of the composed system, a full and atomic invariant which is a Cantor set.

7.4 From Cantor Relation to Truncated Logistic Map

Here, we do not pretend to give a complete study of the logistic map since it is treated in almost every textbook on dynamical systems (e.g. [88, 154]). We analyze the invariant structure of this classical dynamical system as in the two previous cases. However, this analysis slightly differs from the first two ones.

We derive the result from the analysis of the Cantor relation, by giving transformation steps that all preserve the assumptions of union-invariant theorems. Indeed, a very important property in the study of dynamical systems is the stability of systems under small changes or perturbations; structural stability and perturbation theory provide the theoretical foundations of transformations like the ones applied below (see e.g. [88, §1.9] and [264]).

The System. The logistic map is defined on $[0, 1]$ as follows:

$$g(x) = \lambda x(1 - x).$$

When the parameter λ increases, the dynamics of this system varies from very simple (fixpoint attraction, for λ close to 0) to very complex (chaos, for $\lambda \geq 4$).

Let us show how the qualitative compositional analysis of the chaotic truncated logistic map (e.g. with $\lambda = 5$) can be obtained from the Cantor relation.

Transformation Steps. Successive transformations can be applied to the first system, that all preserve initial qualitative properties (see Fig. 7.6, (1)–(4)):

1. starting from the Cantor relation

$$f = \underbrace{(\frac{x}{3})}_{f_0} \cup \underbrace{(\frac{x+2}{3})}_{f_1},$$

2. we first reverse the upper branch

$$\underbrace{(\frac{x}{3})}_{f_0} \cup \underbrace{(1 - \frac{x}{3})}_{f_2};$$

3. then we invert the whole system

$$\underbrace{(3x)}_{f_0^{-1}} \cup \underbrace{(3 - 3x)}_{f_2^{-1}};$$

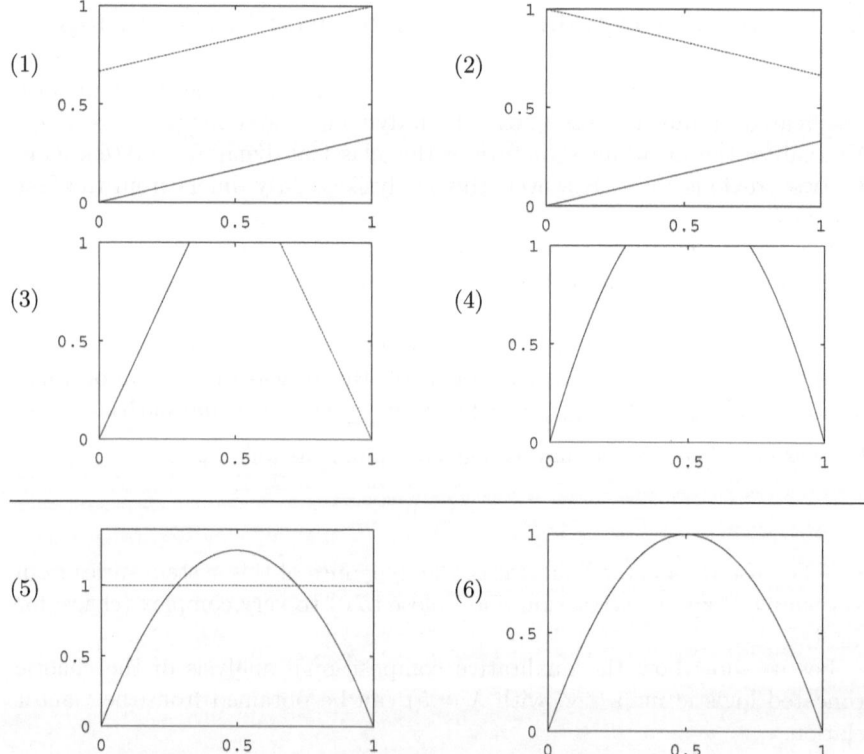

Fig. 7.6. From left to right and top to bottom: (1) Cantor relation, (2) reversing the upper branch, (3) inverting the system, (4) folding and smoothing, (5) zooming (logistic map with $\lambda = 5$); (6) fully chaotic logistic map (with $\lambda = 4$).

4. finally, we fold-and-smooth this piecewise linear relation, and... we get the truncated logistic map

$$g \;=\; g_0 \cup g_1$$

$$g_0(x) \;=\; ([0, \frac{5 - (5)^{\frac{1}{2}}}{10}] \to 5x(1-x))$$

$$g_1(x) \;=\; ([\frac{5 + (5)^{\frac{1}{2}}}{10}, 1] \to 5x(1-x)).$$

Remark 7.1. The last step is not obtained by magic! It suffices to replace x by $\frac{5}{3}x(1-x)$ in f_0^{-1}, and by $1 - \frac{5}{3}x(1-x)$ in f_2^{-1}. Of course, these two transformations add a quadratic aspect to the dynamics.

Validation of the Transformations. Let us now briefly explain why these successive steps preserve the qualitative analysis.

1. Initially, all assumptions of Theorem 6.28 are verified:
   - $([0, 1], \frac{x}{3})$ and $([0, 1], \frac{x+2}{3})$ are contracting with $\gamma = \frac{1}{3}$;
   - $(\frac{x}{3})^{-1}([0, 1]) = (\frac{x+2}{3})^{-1}([0, 1]) = [0, 1]$; $(\frac{x}{3} \cup \frac{x+2}{3})([0, 1]) \subseteq [0, 1]$;
   - each component is injective on $[0, 1]$;
   - $\frac{x}{3}([0, 1]) \cap \frac{x+2}{3}([0, 1]) = \emptyset$.

   The individual invariants are $J^{f_0} = 0$ and $J^{f_1} = 1$. Thus, as we know from §7.3, the original system has a full, atomic, Cantor-set invariant.
2. Changing $\frac{x+2}{3}$ into $1 - \frac{x}{3}$ does not change anything qualitatively, since the contractivity factor γ remains equal to $\frac{1}{3}$.

   Of course, the localization of the resulting Cantor-set invariant varies, due to the new fixpoint invariant of the upper branch: $J^{f_2} = \frac{3}{4}$ (see Prop. 6.31).
3. Inverting the system swaps past and future, but the global invariant is left unchanged. The kind becomes expanding and the contractivity factor becomes $\gamma = 3$.
4. The last step brings slight perturbations. The contractivity factors of both components become $\gamma = (5)^{\frac{1}{2}}$ (see Ex. 5.50). The individual invariants respectively become 0 and $\frac{4}{5}$; these invariants are repulsing fixpoints. Nothing else is changed, and the global analysis remains valid. The last justification holds because the transformations needed to get the quadratic branches from the linear ones are very regular.

Remark 7.2. If we decrease λ down to 4 to get the classical fully chaotic map (see Fig. 7.6, (4)–(6)):

$$4x(1 - x),$$

there is one modification that actually restricts the application of Theorem 6.28.

The first map can be decomposed into two simple variant relations because, on the interval $[0, 1]$, the absolute value of the derivative of the map is always greater than 1; hence, both branches are expanding.

When λ decreases, the central part between $\frac{\lambda-1}{2\lambda}$ and $\frac{\lambda+1}{2\lambda}$ has an absolute derivative smaller than 1. Even if the central part can be cut into two pieces, one for each branch, the theorem cannot be used anymore as such, because there are two different kinds per component. Other methods can be used, like symbolic dynamics, to investigate these limit cases [88].

7.5 Paperfoldings

In this section, we show how our approach can be used to study formal systems. As an example, we concentrate on a formal system describing infinite paperfoldings. Thus, the compositional analysis of dynamical systems does not only apply to classical systems, it also encompasses formal systems and generative or rewriting systems.

7.5.1 Introduction

A paperfolding sequence is the sequence of ridges and valleys obtained by unfolding a sheet of paper which has been folded infinitely many times.

Paperfolding sequences and their complexity have been studied by several authors, using formal power series, continued fractions, language theory and morphisms, measure theory, group theory, etc. [10, 13, 14, 36, 80, 83, 219, 218].

The folding process behind the abstract mathematical terms used to describe these infinite symbolic sequences has been analyzed in another field: hyperbolic dynamical systems [326, 34, 250]. For example, Smale's "stretch-and-squeeze" horseshoe map shows a typical chaotic behavior due to iterative folding of its underlying state space (see §7.2).

Inspired by these results, we propose a way to characterize the complexity of paper folding sequences as in the behavior of horseshoe-like maps and other chaotic dynamical systems: we consider paper foldings as dynamical systems. Up and down foldings correspond to very simple dynamical systems defined on the space of infinite sequences of valleys and ridges. Mixing up and down foldings is shown the be equivalent to composing the corresponding systems in an adequate way. The composed system has an invariant which is the set of all possible sequences. Using composition, we prove that this invariant is a Cantor set, on which the system behaves in a chaotic way. Again, composition is used as a tool to explore the complexity of systems. The approach allows to treat complexity has a structural property of some systems, which avoids long technical developments usually found in classical references.

This section is organized as follows: we first present the formal definition of paperfoldings and paperfolding sequences; we then analyze foldings as dynamical systems, which are found to be chaotic on a Cantor set; finally, we draw some partial conclusions.

7.5.2 Paperfolding Sequences

A *folding action* can be either up (U) or down (D) (see Fig. 7.7); an *instruction* is a sequence of actions; the set of instructions is denoted by $\mathcal{J}$. The

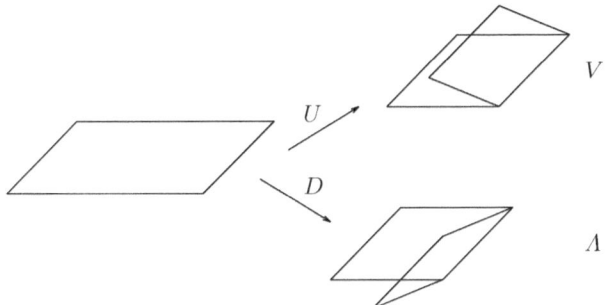

Fig. 7.7. Two possible foldings, up (U) and down (D), and the resulting profiles V and Λ

elementary result of a folding action is a *profile*; it can be either a *valley* (V) or a *ridge* (Λ); a *landscape* is a sequence of profiles; the set of landscapes is denoted by $\mathcal{L}$.

Let $\{U, D\}$ and $\{V, \Lambda\}$ be alphabets, we use the following notations: $\mathcal{J}^{n/*/\omega} = \{U, D\}^{n/*/\omega}$ and $\mathcal{L}^{n/*/\omega} = \{V, \Lambda\}^{n/*/\omega}$.

Not all landscapes are "legal" in the sense they should be obtainable by successive folding actions. Let us give the recursive definition of "legal" landscapes, that is, paperfolding sequences [14].

Definition 7.3 (Paperfolding sequence). *The infinite word* $(w_n)_{n \geq 1} \in \mathcal{L}^\omega$ *is a paperfolding sequence iff* $\forall n \geq 0$

$$w_{4n+1} = V \text{ (resp. } \Lambda)$$
$$w_{4n+3} = \Lambda \text{ (resp. } V)$$

and $(w_{2n})_{n \geq 1}$ *is a paperfolding sequence, too.*

We now turn to the iterative construction of legal landscapes. From an empty landscape ε, i.e. a clean paper, folding up or down leads to a folded paper, nothing else. We must unfold this paper in the reverse order to get a new landscape. Thus, the first point to make precise is what we call an "action" does not really correspond to the folding alone, nor to the unfolding alone, but to both folding then unfolding in the reverse order.

Now, we would like to characterize the semantics of the expressions of the language $\mathcal{J}^*(\mathcal{L}^*)$. Let us first apply U or D to the finite landscape obtained after a finite instruction.

Intuitively (see Fig. 7.8), a folding action consists in inserting between each profile another profile, since all existing ones lie at the borders of the folded paper (B's in the figure) and the folding takes place in the middle of

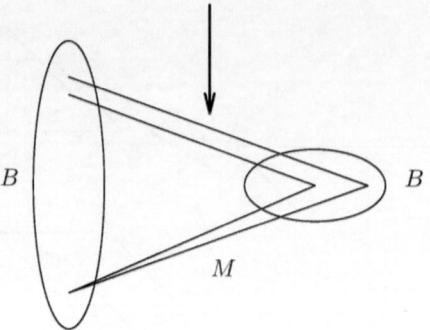

Fig. 7.8. A folded landscape after UD

the folded paper (M in the figure, under the arrow). Of course, the extreme borders do not represent anything in the landscape. Formally, we have the following definition.

Definition 7.4 (Paper folding construction – 1). *Let* $w = w_1 w_2 \cdots w_n$ *be in* $\mathcal{L}^*$, *i.e.* $\forall i, w_i \in \mathcal{L}$; *then*

$$
\begin{aligned}
U(w) &= V & \text{if } w = \varepsilon \\
     &= V w_1 \Lambda w_2 V \cdots V w_n \Lambda & \text{if } n \text{ is odd} \\
     &= V w_1 \Lambda w_2 V \cdots \Lambda w_n V & \text{if } n \text{ is even}; \\
D(w) &= \Lambda & \text{if } w = \varepsilon \\
     &= \Lambda w_1 V w_2 \Lambda \cdots \Lambda w_n V & \text{if } n \text{ is odd} \\
     &= \Lambda w_1 V w_2 \Lambda \cdots V w_n \Lambda & \text{if } n \text{ is even}.
\end{aligned}
$$

Secondly, we extend these definitions by composition to get $\mathcal{J}^*(\mathcal{L}^*)$, i.e. finite instructions applied to finite landscapes.

Definition 7.5 (Paper folding construction – 2). *For* $a \in \{U, D\}, W \in \mathcal{J}^*, w \in \mathcal{L}^*$, *we have:*

$$
\begin{aligned}
\varepsilon(w) &= w \\
aW(w) &= W(a(w)) \\
Wa(w) &= a(W(w)).
\end{aligned}
$$

Remark 7.6. Let us notice that there exists a couple of morphisms u and d related to U and D in the following sense: after a sequence of foldings, the paper is "virtually" folded but what we see is a clean paper (again, see Fig. 7.8); when we unfold this paper in the reverse order, this generates a sequence of landscapes converging to the unfolded landscape given by the previously defined operators.

More formally, let us assume that we execute from a clean paper ε the sequence of actions given by the word W. We call w the resulting landscape, i.e. $w = W(\varepsilon)$. Since $W = W_1 W_2 \cdots W_n \in \mathcal{J}^*$, we rewrite it as $w = W_n(\cdots W_2(W_1(\varepsilon)) \cdots)$. Let us denote the "clean folded paper" as ε'. Unfolding the paper using the mirror image of W gives:

$$\varepsilon' \xrightarrow{\widehat{W_n}} w_n \xrightarrow{\widehat{W_{n-1}}} \cdots \xrightarrow{\widehat{W_1}} w_1$$

where $\widehat{U} = u$, $\widehat{D} = d$, and $w_1 = w$. The morphisms u and d are defined by

$$w \xrightarrow{u} wV\underline{w}$$
$$w \xrightarrow{d} w\Lambda\underline{w}$$

where $\underline{w}$ is defined, for all $w \in \mathcal{L}^*, w' \in \{V, \Lambda\}$, as follows: with $\widehat{V} = \Lambda$ and $\widehat{\Lambda} = V$,

$$\underline{\varepsilon} = \varepsilon$$
$$\underline{w'w} = \underline{w}\widehat{w'}.$$

Intuitively, since u corresponds to one step of unfolding corresponding to an application of U, it is easy to explain the action of u: when a landscape w is up-unfolded, a V appears in the middle, the word w is still on the left, and its reverse mirror image ($\underline{w}$) appears on the right side of the landscape. Successive unfoldings lead to the landscape obtained by a direct application of U.

Finally, extending instructions and landscapes to infinity is straightforward, using the classical continuous limit of finite embedded sequences of increasing length.

A partial order can be defined on words: $\forall u \in \Sigma^*, v \in \Sigma^* \cup \Sigma^\omega, (u \preceq v) \Leftrightarrow (\exists w \in \Sigma^* \cup \Sigma^\omega, uw = v)$. In this case, u is a prefix of v. Moreover, u is a strict prefix of v, $u \prec v$, if $u \neq v$. All finite prefixes of a word constitute a total order. It is also a complete lattice: $(u \sqcup v = u) \Leftrightarrow (u \preceq v) \Leftrightarrow (u \sqcap v = v)$. Thus, every $\preceq$-increasing chain $(w_i)_i$ as a unique least upper bound $\sqcup_i w_i$.

Let us fix $w \in \mathcal{L}^\omega$. For each n, $w|_{2n-1} \prec w|_{2n+1}$ and thus $U(w|_{2n-1}) \prec U(w|_{2n+1})$. Hence, the sequence $(U(w|_{2n+1}))_n$ is strictly $\preceq$-increasing and converges to a unique limit in $\mathcal{L}^\omega$.

Definition 7.7 (Paper folding construction – 3). *Let us fix $w \in \mathcal{L}^\omega$; then*

$$U(w) = \sqcup_n U(w|_{2n+1})$$
$$D(w) = \sqcup_n D(w|_{2n+1})$$

where $\sqcup$ expresses the least upper bound defined by the prefix ordering on sequences.

Extending this to $\mathcal{J}^*$ is straightforward, since every elementary instruction is defined from $\mathcal{L}^\omega$ to $\mathcal{L}^\omega$, and since the same reasoning as above can be applied a finite number of times.

Before characterizing $\mathcal{J}^\omega(\mathcal{L}^\omega)$, it is important to remark that an infinite landscape can only appear after an infinite instruction. Thus, writing $U(w)$, where w is an infinite landscape, is equivalent to $U(W(\varepsilon))$, where W is an infinite instruction leading to the constructible landscape w from the empty landscape, i.e. the clean paper. The last expression can be rewritten as $WU(\varepsilon)$ and it justifies to introduce a right-juxtaposition to infinite instructions. Moreover, thanks to the definitions of U and D given here above, we intuitively see that if we want to find the first letters (that is, the leftmost ones) of an infinite landscape w appearing after an infinite instruction W, it is more useful to know the rightmost part of W than its leftmost part. Symbolically, this reads:

$$W = W_1 W_2 \cdots \quad \Rightarrow \quad WU(w) = U(\underbrace{\cdots W_2(W_1(w)) \cdots}_{?})$$

and

$$W = \cdots W_2 W_1 \quad \Rightarrow \quad WU(w) = U(W_1(W_2(\underbrace{\cdots w \cdots}_{?})))$$

where "?" stands for "not precisely known". The second case is better because the unknown part is less important. For any landscape $w \in \mathcal{L}^\omega$, n folding actions shift w of $2^n - 1$ positions to the right. To know $W(w)$ with a finite precision of $2^n - 1$ profiles, it thus suffices to know the n last actions of W, independently of w. Actually, for $W \in \mathcal{J}^\omega$ and $w \in \mathcal{L}^\omega \cup \mathcal{L}^*$, $W(w) = W(\varepsilon)$ since $2^n - 1$ tends to ω as n does.

Notation 7.8. *From now on, when speaking about infinite instructions, we shall consider infinite words on $\mathcal{J}^\omega$ whose rightmost end is known, and for any $W = \cdots W_2 W_1 \in \mathcal{J}^\omega$, $W|_n = W_n \cdots W_1$, i.e. according to Not. 4.13, $W = W_,$.*

We have $\mathcal{J}^\omega(\mathcal{L}^\omega) = \mathcal{J}^\omega(\varepsilon)$ and we can define the last expressions in the following way.

Definition 7.9 (Paper folding construction – 4). Let $W = \cdots W_2 W_1$ be in $\mathcal{J}^\omega$, then
$$W(\varepsilon) = \bigsqcup_n (W|_n)(\varepsilon).$$

This expression is well defined since, for each n, $(W|_n)(\varepsilon)$ is a strict prefix of $(W|_{n+1})(\varepsilon)$.

> **Remark 7.10.** (See also Rem. 7.6.) Actions u, d extended to infinite words are equivalent to identity. Since we do not treat these functions in the sequel, we just mention this as a remark.

7.5.3 Dynamical Complexity of Paperfoldings

Now, we consider paperfoldings as dynamical systems on symbol sequences, and we characterize the invariants and dynamics of $U \cup D$, and their inverses, using Theorem 6.36.

The functions U and D we use are defined in §7.5.2 (see Def(s) 7.4–7.9). Their domain is the set of infinite landscapes $\mathcal{L}^\omega$.

Metric Properties of Foldings. Let us fix a metric in $\mathcal{L}^\omega$ (it also holds on $\mathcal{J}^\omega$). We consider Def. 2.53 with a parameter c instead of 2^{-1}:

$$d_a(x,y) = c^{\inf\{i \mid x_i \neq y_i\}}.$$

Remark 7.11. Although $c < 1$ is sufficient to keep this distance bounded, we will see in the sequel (see proof of Theorem 7.16) that we may need a stronger condition to prove that foldings are chaotic. Hence, we will consider a parameter $c < \frac{1}{2}$.

Using this metric, it is easy to show that U and D are continuous and contracting: $\forall f \in \{U, D\}$,

$$\forall w, w', d_a(f(w), f(w')) \leq c \cdot d_a(w, w').$$

Since we know that $W(w)$ does not depend on w when $W \in \mathcal{J}^\omega$, it can be interesting to consider the well-defined application

$$\begin{aligned} \chi \quad &: \quad \mathcal{J}^\omega \mapsto \mathcal{L}^\omega \\ \text{s.t.} \quad &\quad \chi(W) = W(\varepsilon). \end{aligned}$$

Proposition 7.12. *The application χ is both continuous and injective.*

Proof. To prove continuity, we have to show that $\forall W, \varepsilon, \exists \delta, \forall W'$,

$$d_a(W, W') \leq \delta \Rightarrow d_a(\chi(W), \chi(W')) \leq \varepsilon.$$

If W and W' are identical up to position n, i.e. $d_a(W, W') \leq c^{n+1}$, $\chi(W)$ and $\chi(W')$ are the same up to position $2^n - 1$, i.e. $d_a(\chi(W), \chi(W')) \leq c^{2^n}$. Thus, ε being fixed, it suffices to take $\delta = c^{1 + \log_2 \log_c \varepsilon}$.

Injectivity is easy to prove. Let us suppose that $W, W' \in \mathcal{J}^\omega$ and they differ from position k, i.e. $\forall i < k, W_i = W'_i$ and $W_k \neq W'_k$. In this case, $\forall i < k, (W|_i)(\varepsilon) = (W'|_i)(\varepsilon)$ but $(W|_k)(\varepsilon) \neq (W'|_k)(\varepsilon)$ from position 2^{k-1}. Thus, we have $W \neq W' \Rightarrow \chi(W) \neq \chi(W')$.

Foldings as Dynamical Systems. Let us summarize the properties of the folding functions U and D.

– Foldings are continuous contracting functions.
– Sequences of landscapes obtained by infinite folding instructions are Cauchy sequences and converge to a limit.

– Moreover, by the contraction mapping theorem, , the successive iterations of these functions, starting from any word, converge to a unique limit which is a fixpoint. If we consider uniform instructions like $UUU\cdots$ (resp. $DDD\cdots$), then, by continuity of foldings, the limits are fixed-points of U (resp. D). Remark that these fixed-points are reachable from any initial landscape.

Cantor Structure of Paperfoldings. The main result of this section follows: we prove that the union of up and down paperfoldings has a full, atomic, invariant which has a Cantor-set structure. It is in itself not surprising but the way we prove it is interesting because we use dynamical systems notions in the context of paperfoldings, that is, formal systems. Before proving the theorem, let us prove three lemmas.

Lemma 7.13. *The functions defined in Def(s) 7.4–7.9 are injective.*

Proof. The two functions U and D are clearly injective: $\forall w, w' \in \mathcal{L}^\omega$, if $\exists k, w_k \neq w'_k$, $(U(w))_{2k+1} \neq (U(w'))_{2k+1}$. Their inverses are also injective when restricted to $V \cdot \Lambda \cdot V \cdot \Lambda \cdots$ and $\Lambda \cdot V \cdot \Lambda \cdot V \cdots$ respectively.

Lemma 7.14. *The dynamics of the systems defined in Def(s) 7.4–7.9 have the same kind, i.e. $\kappa(U) = \kappa(D) = -$, and they are such that $\gamma(U)+\gamma(D) < 1$.*

Proof. The dynamics of these functions are compatible: they are both contracting in the future.

The last argument is more technical: $\gamma(U) = \gamma(D)$, thus we have to show that $\gamma(U) \leq \gamma < \frac{1}{2}$ (which is sufficient to guarantee that $\gamma(U) + \gamma(D) < 1$);

$$\gamma(U) \leq \gamma$$
$$\Leftarrow \quad \sup_{x \neq y} \frac{d_a(U(x), U(y))}{d_a(x, y)} \leq \gamma$$
$$\Leftarrow \quad \sup_{x \neq y} \frac{c^{\inf\{i | (U(x))_i \neq (U(y))_i\}}}{c^{\inf\{i | x_i \neq y_i\}}} \leq \gamma$$
$$\Leftarrow \quad \sup_{x \neq y} \frac{c^{1+\inf\{i | x_i \neq x_i\}}}{c^{\inf\{i | x_i \neq y_i\}}} \leq \gamma$$
$$\Leftarrow \quad c \leq \gamma < \frac{1}{2}.$$

Thus, we have to choose a specific c in order to guarantee this last condition.

Lemma 7.15. *The individual invariants of the systems defined in Def(s) 7.4–7.9 are different fixpoints.*

Proof. The invariants of U and D are different: each application of U (resp. D) inserts a V (resp. Λ) at the left end of the word; at infinity, the fixed points cannot be the same. Moreover, they are fixpoints by contraction (Lemma 7.14).

Theorem 7.16. *The union of paperfolding systems defined in Def(s) 7.4–7.9 generates a full and atomic invariant having a Cantor-set structure.*

Proof. To prove the theorem, we apply Theorems 6.25, 6.26, and 6.27. Thus, we have to verify a few assumptions:

- the functions are injective on $\mathcal{L}^\omega$, by Lemma 7.13;
- the dynamics are compatible and contracting, by Lemma 7.14;
- the individual invariants are different fixpoints, by Lemma 7.15.

In conclusion, we deduce that the invariant of the union of these two systems $U \cup D$ is a full and atomic, totallys disconnected, and perfect, whence it is a Cantor set. This union is interpreted as the set of all possible infinite landscapes resulting from infinite instructions.

Remark 7.17. The same result holds for the union of the inverse systems, $(U \cup D)^{-1}$, since our definitions of invariance and related properties are symmetric in time.

Cantor Structure: The Classical Way. There is a classical way to retrieve the previous result. Let us investigate it and compare it with the compositional approach used in the proof of Theorem 7.16. First, we need the following lemma.

Lemma 7.18. *The invariant of the union of U and D, J, is equivalent to $\chi(\mathcal{J}^\omega)$.*

Proof. The invariant is

$$J^{U \cup D} \quad = \quad \cap_{n \in \mathbb{Z}} (U \cup D)^n (\mathcal{L}^\omega).$$

Since $U^{-1}(\mathcal{L}^\omega) = \mathcal{L}^\omega$ and $D^{-1}(\mathcal{L}^\omega) = \mathcal{L}^\omega$, we have $(U \cup D)^{-1}(\mathcal{L}^\omega) = (U^{-1} \cup D^{-1})(\mathcal{L}^\omega) = U^{-1}(\mathcal{L}^\omega) \cup D^{-1}(\mathcal{L}^\omega) = \mathcal{L}^\omega$. Thus, the invariant can be simplified:

$$J^{U \cup D} \quad = \quad \cap_{n \in \mathbb{N}} (U \cup D)^n (\mathcal{L}^\omega).$$

The union $U \cup D$ is monotonic:

$$X \subseteq Y \quad \Rightarrow \quad (U \cup D)(X) \subseteq (U \cup D)(Y).$$

Moreover, $(U \cup D)(\mathcal{L}^\omega) \subseteq \mathcal{L}^\omega$. Hence, we rewrite the invariant as follows:

$$J^{U \cup D} = \lim_{n \to \infty} (U \cup D)^n (\mathcal{L}^\omega).$$

Finally, $(U \cup D)^n (\mathcal{L}^\omega) = \cup_{w \in \{U,D\}^n} w(\mathcal{L}^\omega)$, and

$$
\begin{aligned}
J^{U \cup D} &= \lim_{n \to \infty} \cup_{w \in \{U,D\}^n} w(\mathcal{L}^\omega) \\
&= \cup_{w \in \{U,D\}^\omega} w(\mathcal{L}^\omega) \\
&= \chi(\mathcal{J}^\omega).
\end{aligned}
$$

Let us now give another proof of Theorem 7.16.

Proof. Since $\mathcal{J}^\omega$ is a Cantor set, and χ is an injective continuous function from $\mathcal{J}^\omega$ to $\mathcal{L}^\omega$, $\chi(\mathcal{J}^\omega)$ is a Cantor set. Thus, J is a Cantor set, too.

Comparison. The classical way involves a quite technical lemma and a proof treating a global system. Our compositional approach states the problem differently: once the system is decomposed into simple subsystems, some easy assumptions have to be verified, and the global result follows automatically. Of course, technically speaking, we have to compare Lemma 7.18 with union-invariant theorems, but the "end-user" can consider their proofs as black boxes. This is the general advantage of any compositional (i.e. modular) approach.

Chaos in Paperfoldings. In addition to the result of Theorem 7.16, it is also possible to show that the paperfolding dynamical system is chaotic on its Cantor-set invariant.

Corollary 7.19. *The dynamical system defined by Def(s) 7.4–7.9 and its inverse are both chaotic on their invariant set.*

Proof. By Theorem 7.16, the invariant of the system is full and atomic. By Prop(s) 5.38 and 5.39, it is thus topologically transitive and sensitive to initial consitions. Hence, by Def. 5.30, the system is Knudsen chaotic.

7.5.4 Partial Conclusions

Studying these paperfoldings as composed dynamical systems is instructive because the approach allows to restate old results in a clear way, like the presence of a Cantor invariant set on which the system is chaotic.

As illustration of union invariant theorems, we have seen that composing two dynamically compatible systems with different fixpoints can lead to a complex behavior sustained by a structurally rich (i.e. Cantor-set structure) invariant set.

We have here a typical example of rich behavior, dynamically complex, resulting from the evolution of a system with a simple structure, the union composition of simple systems.

7.6 Discussion: Compositional Dynamical Complexity

In this chapter, we studied three well-known examples of complex systems by compositional analysis (Smale horseshoe map, Cantor relation, logistic map). This led to rederive classical results very clearly, namely the Cantor-set structure of the invariants of these systems. It is up the the reader to judge whether this rederivation is clearer or not than the classical approaches [292, 293, 326, 88].

Then, we embedded paperfoldings in the context of dynamical systems and we showed that these systems are chaotic on a Cantor invariant set, using a straightforward decomposition of a global system into subsystems respectively corresponding to up and down foldings (Theorem 7.16). Considering formal systems as dynamical systems seems thus a fruitful and promising approach [268].

Thus, the same phenomenon has been observed in different kinds of systems like classical dynamical systems (e.g. Smale Horseshoe Map, Cantor relation, logistic map), symbolic systems (e.g. paperfoldings): complexity arises from the composition of compatible systems attracting the space to different regions in the future or in the past. A kind of hyperbolic behavior sustains all these rich dynamics in complex systems composed from simpler ones having simple, attracting, symmetric dynamics.

8. Experimental Compositional Analysis of Cellular Automata

Cellular automata (for short, CA) are *massively parallel systems* obtained by composition of myriads of *simple agents* interacting locally, i.e. with their closest neighbors. In spite of their simplicity, the dynamics of CA is potentially very rich, and ranges from attracting stable configurations to spatiotemporally chaotic features and pseudo-random generation abilities, from very simple forms of destruction of information to more complex ones where information propagates following non-trivial rules [4, 249, 107, 113, 115, 121, 132, 291, 317, 331]; Von Neumann introduced them in order to model biological self-reproducing behaviors [314]. Moreover, from the computational viewpoint, they are universal, that is, as powerful as Turing machines and, thus, classical Von Neumann architectures (see Chap. 9). This motivates our choice to study these highly stuctured systems in more details; we concentrate on two aspects.

First, we establish a formal classification of behaviors based on attraction, inspired by the phenomenological classification schemes proposed in [330, 49]. This allows us to propose a view of dynamical complexity for these systems (see also §5.7.3), and brings new insights in understanding complexity of high-dimensional systems.

Second, we present compositional arguments to evidence a conjecture on the emergence of complexity from the composition of elementary symmetric compatible systems [49, 101]. This second part is based on the union-invariant theorems elaborated in Chap. 6, that we complement with physical measures of complexity; this serves to cope with the difficult analysis of the underlying connected product.

The present chapter is organized as follows. In §8.1, we motivate the topics treated in the chapter. Then, we develop an attraction-based classification: in §8.2, we define preliminary notions, as well as the tools used to classify behaviors; in §8.3, we present a phenomenological attraction-based classification of behaviors, which is then formalized in §8.4; three structural organizations of classes are proposed in §8.5, in order to understand dynamical complexity in cellular automata. We proceed to the compositional analysis of CA: in §8.6, we give experimental conjectures on the disjunctive composition of cellular automata; in §8.7, we compare a particular disjunction to the Cantor relation; in §8.8, we define the successive steps of our analysis and we add complexity

F. Geurts: Abstract Compositional Analysis of Iterated Relations, LNCS 1426, pp. 183-216, 1998.
© Springer-Verlag Berlin Heidelberg 1998

measures to our theoretical framework; in §8.9, we analyze the conjectures. Finally, we close the chapter with a discussion in §8.10.

8.1 Aims and Motivations: Attraction-Based Classification and Composition

On one hand, the structural and dynamical features of CA make them very powerful: fast CA-based algorithms are developed to solve engineering problems in cryptography and microelectronics for instance (e.g., [4, 55, 54, 59, 249, 128, 135, 239, 281, 291]), and CA-based models are used in ecology, biology, physics, telecommunications, and image-processing (e.g., [113, 115, 132, 263]).

On the other hand, these powerful features make CA difficult to analyze: they resist against most of the analytical tools available in program theory and dynamical system theory. Actually, almost all long-term behavioral properties of dynamical systems, and cellular automata in particular, are unpredictable. This limit not only arises from their possibly chaotic dynamics, which entails the classical sensitivity to initial conditions, but from a much stronger limit: undecidability [74, 75, 224]. Sensitivity means that close initial conditions eventually diverge. Undecidabilty implies that no algorithm exists whatsoever to decide whether or not an arbitrary state is attracted to some set, given the general description of some dynamical system.

Two pragmatic solutions exist to this problem: the first one consists in restricting the study to particular cases and sufficiently simple systems that may be used as building blocks; the second one is to make an extensive use of computer simulations. Both solutions are necessary engineering tools in the construction of systems that must be built in order to fulfill a priori conditions or specifications.

In the following sections, our objective is the study of dynamical properties of CA-based systems. The main property we focus on is attraction, as many behaviors of systems can be rephrased in these terms. However, due to the severe theoretic limitations explained above, we carry out an *experimental analysis* of systems in a *compositional* way, i.e. by combination of properties of their components: we analyze attraction properties of basis CA, and we combine these basis CA to obtain the dynamics of a whole family of CA-based systems.

In particular, our aim is to make use of the very nature of the structure of CA to understand their global behavior, since dynamical attraction-based properties of CA apparently depend more strongly on their structure than on the way initial conditions are chosen. Given two elementary CA (Boolean one-dimensional bi-infinite lattices of cells, the evolution of each cell being influenced by its direct neighbors) with known behaviors, what can be said on the behavior of the CA obtained by composing them (e.g., using a logic disjunction)?

Classification. In the theory of CA, classification of behaviors is a central theme [132, 113]. The goal is to impose a structure in the space of CA rules, grouping together CA related to equivalent properties. Up to now, different tools have been introduced, leading to different classification schemes.

Two problems appear in many classification schemes. First, some classifications happen to be informally defined (e.g. [330]). Second, classification schemes are sometimes far from the intuition we get by observing the behavior of CA classes (e.g. [41]).

We classify the attraction properties of cellular automata into periodic, shifting, and aperiodic or complex behaviors. Then, we propose structural organizations of the resulting classes, which allows us to clearly separate shifting behaviors from complex ones.

The goal is not to propose an algorithm to decide which class a given cellular automaton can belong to, because this is undecidable in general. We rather examine a classification scheme that better corresponds to the qualitative classification we can arrive at when observing the evolution of cellular automata, in order to understand dynamical complexity of structured high-dimensional systems showing spatiotemporally chaotic behaviors.

Composition. CA are adequately modeled by connected products, as mentioned in Ex. 3.24: any CA can be written as $\otimes_R g$, where g is a local rule distributed over a regular lattice, and R describes interactions between neighbors.

In [49], the authors made the conjecture that that the disjunction of some local rules entails complex behaviors . The second aim of this chapter is to analyze this conjecture by composition: we relate these experimental results to the compositional analysis of unions of systems.

The previous chapters have shown that the general analysis of connected products is rather difficult (in particular, see §§3.3, 6.8). Here, we focus on a specific family of connected products and we introduce complexity measures which complement our previous compositional results.

Composing CA can be realized in two ways: either *globally*, i.e. considering global functions as black boxes, or *locally*, i.e. at the level of local transition functions. More formally, let $\star$ and $\star'$ be two composition operators. The composition of several CA $\otimes_R g_i$ can be defined outside the connected products (this is the usual global manner), or inside the product:

$$\star_i(\otimes_R g_i) \text{ or } \otimes_R (\star'_i g_i).$$

From an engineering viewpoint, local functions are building blocks, and they are supposed to be well understood, whereas global functions are to be built, and composing global functions is beyond the scope of this paper. Hence, our concern here is the study of local compositions composed within the connected product: f and g being two local transition functions, $\star$ being a local composition operator, we study the system $\otimes(f \star g)$. If G is a global property, and I an individual property, the objective is to find $\diamond$ such that

$$G(\otimes_R(\star'_i g_i)) = \diamond_i I(\otimes_R g_i).$$

However, this kind of local composition clearly adds a difficulty to the analysis of systems, because the property composition operator $\diamond$ has to "jump" over the connected product. This is equivalent to finding intermediate G' and $\diamond'$ such that

$$G(\otimes_R(\star'_i g_i)) = G'(\diamond'_i(\otimes_R g_i)) = \diamond_i I(\otimes_R g_i).$$

In the particular cases studied here, local rules are decomposed so that the previous equalities hold regarding complexity measures: the global property G and the individual property I both correspond to a complexity level in the classification introduced. Again, we show how the composition of simple systems generates complex behaviors.

8.2 Preliminary Notions

In this section, we briefly recall the definition of cellular automata. Then, we introduce two tools used to classify attracting behaviors of CA and to organize the structure of the resulting classes.

8.2.1 Cellular Automata

CA are totally discrete dynamical systems. They are discrete *in space*; in fact, they are regular lattices of sites (or cells) whose values range in a finite set. They are also discrete *in time*: their iterative dynamics is described in terms of difference equations, as opposed to continuous-time systems based on differential equations. Furthermore, CA are homogenenous systems: at each time step, each cell applies the same rule to its neighborhood in order to compute its new value.

We consider one-dimensional CA, i.e. linear bi-infinite lattices of cells. Each cell takes its value in the local state space $X = \{0, 1, \ldots, k - 1\}$. A *configuration* is a bi-infinite sequence of $X^{\mathbb{Z}}$ specifying a state for each cell, i.e. $x = (\cdots x_{-1}, x_0, x_1 \cdots)$ (see Fig. 8.1). The *neighborhood* of a cell $i \in \mathbb{Z}$ is $(i - r, \ldots, i - 1, i, i + 1, \ldots, i + r)$ or, simply, $(i - r : i + r)$.

$$\cdots \leftrightarrow \boxed{x_{-2}}_{-2} \leftrightarrow \boxed{x_{-1}}_{-1} \leftrightarrow \boxed{x_0}_0 \leftrightarrow \boxed{x_1}_1 \leftrightarrow \boxed{x_2}_2 \leftrightarrow \cdots$$

Fig. 8.1. Cellular automaton: x is the global configuration, the numbered boxes represent cells, and each x_i is a corresponding local cell value.

All cells are updated synchronously. At each step, every cell looks at the value of its neighbors (r to the left, r to the right) plus itself and computes its next value as a function of this neighborhood. This function is a *local transition function* $g : X^{2r+1} \mapsto X$. There are clearly $k^{k^{2r+1}}$ different local functions.

The *global transition function* defines the next state of each cell as the local function applied to the states of its neighborhood:

$$f \quad : \quad X^{\mathbb{Z}} \mapsto X^{\mathbb{Z}}$$
$$\text{s.t.} \quad \forall i \in \mathbb{Z}, f_i(x) = g(x_{i-r:i+r}).$$

In the following, we restrict our attention to *elementary cellular automata*, i.e. with $r = 1$ and $k = 2$. With these parameters, there are 256 different elementary cellular automata. Each transition function is a Boolean function of three Boolean variables (the neighborhood) and is thus expressed as a *transition (or rule) table* with eight entries. Traditionally [330], a unique integer is associated to each transition function, and is used to label the corresponding CA:

$$\sum_{a,b,c \in \{0,1\}} g(a,b,c) \cdot 2^{4a+2b+c}.$$

Example 8.1. The following table corresponds to rule $47 = 1 + 2 + 4 + 8 + 32$.

$x_{i-1}^t x_i^t x_{i+1}^t$	000	001	010	011	100	101	110	111
x_i^{t+1}	1	1	1	1	0	1	0	0

A local state x is *quiescent* iff $f(\cdots xxx \cdots) = \cdots xxx \cdots$ or $g(\underbrace{x \cdots x}_{2r+1 \text{ times}}) = x$.

Finally, using the operators on systems introduced in Chap. 3, we can express CA in the following way (see also Ex. 3.24).

Definition 8.2 (Cellular automaton). *A cellular automaton f is structured as follows:*

$$f = \otimes_R g$$

where

- $J = \mathbb{Z}$ *is the lattice of cells;*
- $R = \{(i, i-1), (i,i), (i, i+1) \mid i \in J\}$ *describes the neighborhood of each cell;*
- $\forall i \in J, X_i = \{0,1\}$ *is the local state space;*
- $\forall i \in J, g_i = g : X^3 \mapsto X$ *is the local transition function.*

Remark 8.3. Let us notice that we use a configuration space $\mathcal{C}$ the cardinality of which is $2^{\aleph_0}$, at least equal to $\aleph_1$. If we consider the power set of this configuration space, i.e. the set of all subsets of $\mathcal{C}$, the cardinality is $2^{\aleph_1}$, at least equal to $\aleph_2$. This cardinality is equal to the one of the power set of an interval of $\mathbb{R}$. We know that $\aleph_1, \aleph_2, \aleph_3, \ldots$ are greater than the first transfinite ordinal number ω and we will see below why transfinite iterations are useful in this framework.

8.2.2 Transfinite Attraction

Working in a space E, we consider the set of its subsets, $\mathbb{P}(E)$. It is well known that this set is a complete lattice. We denote it as follows: $\mathbb{P}(E)(\subseteq, \emptyset, E, \cap, \cup)$

Using transfinite iterations (see Def. 2.76), it is possible to refine Def. 5.51. This gives the following definition.

Definition 8.4 (Transfinite attraction). *P is transfinitely attracted by Q iff there exists an ordinal number $n \in \mathbb{O}$ such that, $\forall m \geq n, f^m(P) \subseteq Q$.*

We can restrict ourselves to the smallest Q attracting P. In the following, some notations will be useful.

Notation 8.5. *Strict transfinite (resp. unbounded finite) attraction is denoted by $P \overset{f}{\leadsto}_\eta Q$ (resp. $P \overset{f}{\leadsto}_\omega Q$).*

To find the whole attractor of configuration space $\mathcal{C}$, we have to compute an invariant of the system, $f^n(\mathcal{C})$, for a certain ordinal number $n \in \mathbb{O}$. This expression is computable by successive approximations, and leads to the attractor, thanks to monotonicity of f.

Remark 8.6. The first transfinite ordinal number is large enough to compute the global attractor, since every CA is continuous (see Chap. 9) and continuity implies convergence in at most ω steps.

8.2.3 Shifted Hamming Distance

We introduce here a notion of distance that is very close to the well-known Hamming distance. Let us first extend this notion, defined on finite strings of symbols, to bi-infinite strings of symbols.

We work with a finite alphabet $\Sigma = \{0, 1, ..., k-1\} \subseteq \mathbb{N}$. On this alphabet we define a distance

$$d_s(x, y) = \begin{cases} 0 & \text{if} \quad x = y \\ 1 & \text{if} \quad x \neq y \end{cases}$$

For two strings of symbols $a, b \in \Sigma^m$, the *Hamming distance* between a and b, $H(a, b)$ is defined as the number of places (or indices) where a and b differ:

$$H(a, b) = \sum_{i=1}^{m} d_s(a_i, b_i).$$

For two bi-infinite sequences a and b of symbols,

$$H(a, b) = \sum_{i \in \mathbb{Z}} d_s(a_i, b_i).$$

Definition 8.7 (Shifted Hamming distance). *The shifted Hamming distance between two bi-infinite sequences x and y of $\mathcal{C}$ is defined by:*

$$H^\rho(x, y) = \min_{j \in \mathbb{Z}} \upsilon(H(x, \rho^j(y)))$$

where ρ is the classical shift: $\forall i \in \mathbb{Z}, \rho(x)_i = x_{i+1}$, and $\upsilon(x)$ is defined as follows:

$$\upsilon(x) = \begin{cases} \frac{1-e^{-x}}{1+e^{-x}} & \text{if} \quad x \in \mathbb{R}, \\ 1 & \text{if} \quad x = +\omega \end{cases}$$

Remark 8.8. The shift-invariant function defined here is not a metric on $\mathcal{C}$, and it generates a trivial topology on the quotient space $\mathcal{C}/_{\equiv_\rho}$. Motivated by the same need to investigate the meaning of sensitivity and its implications in CA behaviors, a group of researchers recently introduced a shift-invariant pseudo-metric on $\mathcal{C}$ inducing a non-trivial topology [51, 52]. Their metric is based on weighted local Hamming distances, and is thus sustained by the same underlying idea as the shifted Hamming distance. We conjecture that its use in our context would lead to the same results. We leave this comparison to the reader.

8.3 Experimental Classification

Now, we propose a classification of long-term behaviors of CA when they start from random initial conditions. We observe the results of simulations to determine six phenomenological classes grouped into three families: periodic (types $\mathcal{N}$, $\mathcal{F}$, $\mathcal{P}$), shifting (type $\mathcal{S}$, divided into types $\mathcal{S}_\mathcal{F}$ and $\mathcal{S}_\mathcal{P}$), and aperiodic (type $\mathcal{A}$) behaviors. Here is the classification in brief, essentially based on [330, 49, 102]; typical evolutions are illustrated in Fig. 8.2.

Type $\mathcal{N}$: CA evolving to null configurations. This class contains CA that quickly evolve to homogeneous configurations, i.e. without information (all ones or zeroes), after finite transients.

Type $\mathcal{F}$: CA evolving to fixed points. This second class contains CA that evolve to fixed points after finite transients. Of course, class $\mathcal{N}$ is a particular case of class $\mathcal{F}$.

Type $\mathcal{P}$: CA with periodic behaviors. This class contains CA that evolve to periodic configurations, after finite transients. It contains the two previous ones.

Type $\mathcal{S}$: CA with generalized subshift behaviors. This class contains CA that, starting from an initial configuration in particular subspaces of the configuration space, evolve to configurations where a generalized alternating subshift behavior occurs. Here is a definition of this behavior, generalizing [48].

Definition 8.9 (Generalized alternating subshift). *A CA is a generalized alternating subshift rule if the corresponding global function f is such that there is a closed invariant subset Σ_1 of C such that*

$$\forall x \in \Sigma_1, f^n(x) = \rho^m(x)$$

where $n \in \mathbb{N}$ and $m \in \mathbb{Z}$, and $\rho : C \mapsto C$ is the classical shift.

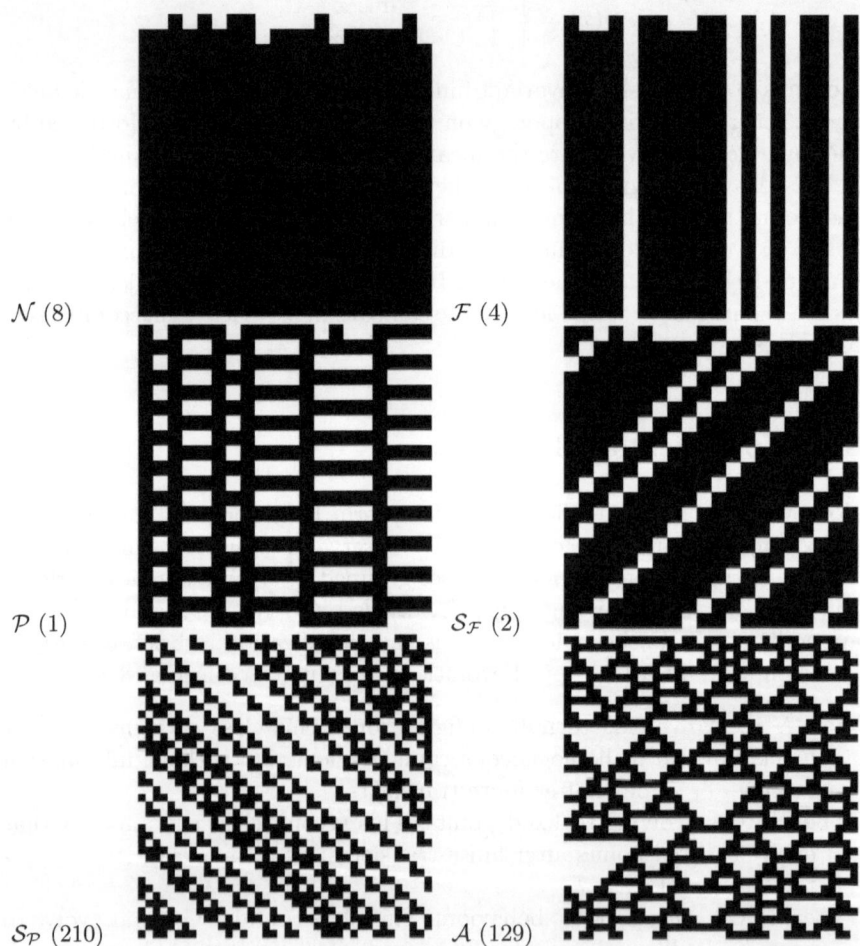

Fig. 8.2. CA classification: typical CA evolutions of classes $\mathcal{N}$, $\mathcal{F}$, $\mathcal{P}$, $\mathcal{S_F}$, $\mathcal{S_P}$, and $\mathcal{A}$. In this figure, the horizontal axis represents a portion of the CA state space, the vertical axis represents the temporal evolution (top-down); black dots are 0's and white dots are 1's.

Remark 8.10. − On the complement of the subspace Σ_1 where the CA acts like a shift, the behavior can be regular or totally irregular.

– As shown in Fig. 8.2, the family of shifting behaviors contains a broad range of behaviors, from very simple ones like the dynamics of rule 2, to very complex ones like the dynamics of rules 210. This is the reason why we distinguish the simplest ones, for which $n = m = 1$ in the above definition (they are denoted by $\mathcal{S}_{\mathcal{F}}$), from the other ones (denoted by $\mathcal{S}_{\mathcal{P}}$). Of course, $\mathcal{S} = \mathcal{S}_{\mathcal{F}} \cup \mathcal{S}_{\mathcal{P}}$. In the following, we will distinguish these two families only when necessary.

Type $\mathcal{A}$: CA with complex or aperiodic behaviors. A configuration is aperiodic if it is not eventually periodic (neither periodic nor one of its forward iterations). Qualitatively, what we observe is a number of different patterns growing, vanishing and moving toward the future. In general, a broad range of behaviors can show up: from random noise, total disorder, and spatiotemporal chaos to some kind of regularity or intermittency in which diverse forms can propagate. Aperiodicity entails that almost the whole domain is visited through successive iterations.

Example 8.11. The initial pattern can evolve smoothly, like a ball flying between two walls, one of the walls being fixed, the other one escaping to infinity as the ball bounces on it (see Fig. 8.3, where the local space $X = \{\cdot, 1, 2, 3, 4, 5\}$).

Transition table				Example of evolution
x^t_{i-1}	x^t_i	x^t_{i+1}	x^{t+1}_i	
$\cdot$	$\cdot$	4	4	.....1....3..2.....
$\cdot$	$\cdot$	5	4	.....1.....3.2.....
$\cdot$	3	$\cdot$	$\cdot$	.....1......32.....
$\cdot$	3	2	$\cdot$	.....1.......5.....
$\cdot$	4	$\cdot$	$\cdot$	.....1......4.2....
$\cdot$	5	$\cdot$	$\cdot$	.....1.....4..2....
1	$\cdot$	4	3	.....1....4...2....
1	3	$\cdot$	$\cdot$	.....1..4....2....
1	4	$\cdot$	3	.....1..4.....2....
3	$\cdot$	$\cdot$	3	.....1.4......2....
3	$\cdot$	2	3	.....13.......2....
3	2	$\cdot$	5	.....1.3......2....
5	$\cdot$	$\cdot$	2	.....1..3.....2....
$\cdot$	b	$\cdot$	b	

Fig. 8.3. Flying ball. In the transition table, $(., b, .) \rightarrow b$ stands for the "else" case.

8.4 Formal Attraction-Based Classification

In §8.3, we presented a phenomenological classification of CA long-term behaviors. This observational point of view has to be kept in mind when devel-

oping a theory of dynamical systems, in order to remain as close as possible
to the intuition we can get by observing the evolution of systems. Based on
this informal classification, we develop a formal classification of CA which
is based on attraction properties. This is the aim of this section: after a few
introductory comments, we systematically present each class of behavior as
stated in §8.3, and we give some remarks.

8.4.1 Introduction

Our motivation to classify behaviors w.r.t. attraction is the following. There
seems to exist three kinds of observable behaviors:

- regular (null, fixed-point, or periodic rules),
- irregular (our type $\mathcal{A}$),
- intermediate behaviors (subshift rules).

The first kind is easily understood. Let us turn our attention to the two
other classes. We see at least three different ways to reduce their noticeable
disorder:

- using transfinite attraction;
  Let us justify this intuitively. We consider a single deterministic automaton with a finite number of possible states. We let the system progress, or
  iterate, and we look at the orbits generated from different initial states. If
  we only take an orbit passing through less states than the total number of
  states in the space, then different behaviors are observable: fixed-point attraction, periodicity, seemingly random orbits. (What random really means
  here is another question [57, 58] and that is why we add "seemingly".) If
  we consider more iterations than the number of possible states, then random orbits disappear since the system is deterministic. Everything becomes
  eventually fixed or periodic. We consider now larger and larger state spaces,
  until we reach infinity. For example, we take a state space of positive inte-
  gers $\mathbb{N}$ of cardinality $\aleph_0$, the first transfinite cardinal, also equal to ω. The
  same behaviors appear and we have to allow more than ω iterations to see
  only periodic behaviors (with possibly huge periods).
- forgetting the origin of the lattice;
  It is interesting to work in $\mathcal{C}/_{\equiv_\rho}$, the quotient space containing equivalence
  classes w.r.t. shift. Indeed, any shifting behavior becomes periodic or even
  fixed in this space.
- increasing the dimension of the observation space.
  For example, it is possible to work with the Cartesian product space $\times$
  time [27, 259, 260], or to embed the space and the attractor of a system in
  a higher-dimensional space, in order to "unfold" its internal interactions
  (this is called "dimension embedding") [250].

In the following, we only consider the first two approaches. The main part of this section treats the first case. In §8.5, we present the second one using the shifted Hamming distance introduced before.

Several choices are possible for the study of attraction phenomena in cellular automata. Some authors work with finite configurations in zero backgrounds (a finite number of cells are initialized at random, all others are set to zero). We consider bi-infinite configurations in zero backgrounds. Since we consider bi-infinite configurations, we have to consider transfinite iterations; then we have to specify what happens at the "borders" of these lattices. Another possible choice is a circular bi-infinite CA, having cells $-\infty$ and ∞ equivalent. We take one configuration at a time ($P = \{x_0\}$), randomly, and we "observe" its successive iterations, over a certain amount of time (finite, infinite, or transfinite): for a cellular automaton the global function of which is given by f, we observe, $\forall n \in \mathbb{O}, f^n(P)$. We also get information by studying the same expression with $P = \mathcal{C}$ (the whole configuration space). Finally, it is interesting to compare the results obtained with or without the constraint of finite iterations.

8.4.2 Type-$\mathcal{N}$ Cellular Automata

This first class contains CA that quickly evolve to homogeneous configurations, i.e. any configuration is finitely attracted to the same configuration, homogeneously composed of quiescent cell states. The homogeneous state is a function of the rule itself:

$$\exists h \in \mathcal{C}, \left\{ \begin{array}{l} \forall x_0 \in \mathcal{C}, \exists n \in \mathbb{N}, f^n(x_0) = h \\ \wedge \quad f(h) = h \end{array} \right.$$

or, more globally,

$$\boxed{\mathcal{C} \rightsquigarrow^f_\omega \{h\}}.$$

This class is called $\mathcal{N}_0$ because another version is possible, with possibly several quiescent configurations:

$$\exists H \subset \mathcal{C}, \forall x_0 \in \mathcal{C}, \exists h \in H, \left\{ \begin{array}{l} \exists n \in \mathbb{N}, f^n(x_0) = h \\ \wedge \quad f(h) = h \end{array} \right.$$

or, more globally,

$$\boxed{\mathcal{C} \rightsquigarrow^f_\omega H}.$$

We call this subclass $\mathcal{N}_1$.

8.4.3 Type-$\mathcal{F}$ Cellular Automata

This class contains CA that evolve to fixed configurations after finite transients. The final fixed configuration is in general dependent on the initial one. We have here a finite attraction, too:

$$\forall x_0 \in \mathcal{C}, \exists s_{x_0} \in \mathcal{C}, \left\{ \begin{array}{l} \exists n \in \mathbb{N}, f^n(x_0) = s_{x_0} \\ \wedge \quad f(s_{x_0}) = s_{x_0}. \end{array} \right.$$

More globally, we have:

$$\boxed{\mathcal{C} \overset{f}{\rightsquigarrow}_\omega \cup_{x_0 \in \mathcal{C}}\{s_{x_0}\}}.$$

8.4.4 Type-$\mathcal{P}$ Cellular Automata

This class contains CA that evolve to cycles of configurations after finite transients. The limit cycle is dependent on the initial condition. We have a finite attraction to a set of points rather than to a single fixed-point:

$$\forall x_0 \in \mathcal{C}, \exists C_{x_0} \subseteq \mathcal{C}, y \in C_{x_0}, m \in \mathbb{N}, n \in \mathbb{N}, \left\{ \begin{array}{l} f^n(x_0) = y \\ \wedge \quad \forall y' \in C_{x_0}, f^m(y') = y' \\ \wedge \quad \forall m' < m, f^{m'}(y') \neq y'. \end{array} \right.$$

Using a more compact notation, we have:

$$\boxed{\mathcal{C} \overset{f}{\rightsquigarrow}_\omega \cup_{x_0 \in \mathcal{C}} C_{x_0}}.$$

8.4.5 Type-$\mathcal{S}$ Cellular Automata

This class contains CA that behave like generalized alternating subshifts. There exists a closed subspace of the configuration space, such that, when observing a specific cellular automaton starting from a random initial configuration in that subspace, what we see is the initial configuration progressively escaping (or shifting) to the right or to the left, like sliding along the linear lattice of cells, together with a kind of periodic behavior. If we take an initial finite configuration in a zero background, for example, we will see our configuration escaping the finite observation domain, unless this domain can grow indefinitely.

Let us imagine we could iterate more than the total amount of cells composing the lattice of the automaton, even if this lattice possesses a bi-infinite number of cells. Then, starting from any initial configuration, we could observe an attraction to a homogeneous configuration, exactly as type-$\mathcal{N}$ cellular automata behave. From this point of view, the behavior of type-$\mathcal{S}$ cellular automata becomes more regular and simpler than chaotic, if we use transfinite iterations. Formally, $\mathcal{C}_f$ being the set of finite configurations in a bi-infinite zero background, we have the following characterization:

$$\exists h \in \mathcal{C}, \left\{ \begin{array}{l} \forall x_0 \in \mathcal{C}_f, \exists n \in \mathbb{O} \backslash \mathbb{N}, f^n(x_0) = h \\ \wedge \quad f(h) = h \end{array} \right.$$

or, more generally,

$$\exists H \subseteq \mathcal{C}, \forall x_0 \in \mathcal{C}_f, \exists m \in \mathbb{N}, n \in \mathbb{O} \backslash \mathbb{N}, \left\{ \begin{array}{l} \exists y \in H, f^n(x_0) = y \\ \wedge \quad \forall y' \in H, f^m(y') = y' \\ \wedge \quad \forall m' < m, f^{m'}(y') \neq y' \end{array} \right.$$

where H is a cycle of homogeneous configurations. It is also possible to write:

$$\boxed{\mathcal{C}_f \rightsquigarrow^f_\eta H}.$$

Remark 8.12. – Here, we see a difference regarding the use of finite/transfinite iterations. Finite iterations lead to a typical shift behavior which can be seen as chaotic (classical definition). Transfinite iterations show a simple behavior of attraction to homogeneous configurations.
– We have treated the case of finite configurations in a bi-infinite zero background, that can be precisely defined. Treating the general case is left as further work.

8.4.6 Type-$\mathcal{A}$ Cellular Automata

This last class contains CA that have an aperiodic behavior which is responsible for the observable (spatiotemporal) chaos. The definition we propose below is quite general. A part of our future work is to find a subclassification with more appropriate and refined definitions. Back to attraction, we have here a "convergence" to a huge cycle containing (almost) the whole configuration space:

$$\forall x_0 \in \mathcal{C}, \exists C' \subseteq \mathcal{C}, m \in \mathbb{O}, n \in \mathbb{O} \backslash \mathbb{N}, \left\{ \begin{array}{l} \exists y \in C', f^n(x_0) = y \\ \wedge \quad \forall y' \in C', f^m(y') = y' \\ \wedge \quad \forall m' < m, f^{m'}(y') \neq y'. \end{array} \right.$$

This inclusion is always strict, Chaitin's theorem saying that most strings, or real numbers, are not computable (reachable by an algorithm) [45, 58]. It is also possible to write:

$$\boxed{\mathcal{C} \rightsquigarrow^f_\eta C'}.$$

Remark 8.13. Here also, we have a difference between finite and transfinite iterations. Finite iterations show irregular behaviors, spatiotemporally chaotic patterns, aperiodic evolutions. The problem is that it is difficult to give an explicit characterization of this kind of behavior. On the other hand, transfinite iterations allow us to give a very simple definition, saying that the system involved is periodic with a huge period very close to the cardinality of the configuration space itself.

8.4.7 Discussion

Type-$\mathcal{N}_1$ CA are not exactly the same as type-$\mathcal{F}$ CA because the latter are dependent on the initial conditions whereas the former are not as much dependent (there is only a small number of possible outcomes). Where the precise border lies between these two classes has not been precisely defined yet. Therefore, for the sake of simplicity, we include $\mathcal{N}_1$ in $\mathcal{F}$ and keep $\mathcal{N}$ equal to $\mathcal{N}_0$.

We see here that transfinite attraction gives us a new way of defining the behavior of differents classes of CA, from very simple classes to the most complex ones: all classes can be seen as periodic, from small finite periods to huge transfinite ones.

If we restrict our attention to basis CA (i.e. rules 0, 1, 2, 4, 8, 16, 32, 64, and 128), our classification is of course decidable. Our goal is to extend the notions and our classification to more complex CA, constructed from basis CA with composition operators (see §§8.6–8.9).

8.5 Structural Organizations of CA Classes

The previous section (§8.4) was devoted to the formal definition of attraction-based classes of cellular automata corresponding to phenomenological classes presented in §8.3. We identified six categories of behaviors: null, fixed, periodic, (fixed and periodic) shifting, and aperiodic or complex behaviors, grouped into three families (periodic, shifting, aperiodic).

In this section, we propose three different but complementary ways to organize and structure these classes into groups to emphasize the regularity or the disorder introduced by each class. The last organization explicitly uses shifted Hamming distance to reduce the disorder of classes, as stated in the introduction of §8.4. With start with a discussion motivating these structural organizations by an interesting paradox, and close the section with a definition of dynamical complexity in CA.

8.5.1 Motivation: Simulation vs Theoretical Results

Let us motivate the need for structural organizations of CA classes using an interesting paradox, where observations and mathematics totally disagree on the interpretation of some behaviors.

Intuitively, periodic behaviors (null, fixed and periodic) are regular, aperiodic behaviors show rather complex evolutions, and shifting behaviors are usually not very complicated, but some initial conditions can lead them to subspaces where nothing can be said.

The paradox we would like to illustrate here precisely concerns shifting behaviors. In spite of the regular behavior depicted in Fig. 8.2, they are Devaney chaotic (Def. 5.28)! The following theorem is inspired by [41], where it was proved for three particular cases ($n = m = 1$; $n = 2$, $m = 1$; $n = m = 2$).

Theorem 8.14. *If Σ_1 is a closed invariant subset of $\mathcal{C}$, and if a CA f is a generalized alternating subshift on Σ_1 with irreducible transition matrix, then it is Devaney chaotic on Σ_1.*

Proof. We have to prove that (see Def. 5.28):

1. there exists a dense orbit for f (leading to topological transitivity),
   It is easy to construct a dense sequence for the full shift. This sequence can be expressed as a sequence of all sequences of length 1, all sequences of length 2, etc. All these sequences can be ordered. We denote the resulting sequence by $\cdots s_5 s_3 s_1 s_0 s_2 s_4 s_6 \cdots$.
   From this, we deduce that ρ^m has a dense orbit, too. The resulting sequence is $\cdots s_5^m s_3^m s_1^m s_0^m s_2^m s_4^m \cdots$.
   Thus f^n has a dense orbit. Hence f has a dense orbit (the same as for f^n).
   All these sequences can be adapted to the case of a subshift with irreducible transition matrix: the final resulting sequence is $\cdots t_3 (s_1 s_1')^{m-1} s_1 t_1 (s_0 s_0')^{m-1} s_0 t_0 (s_2 s_2')^{m-1} s_2 t_2 \cdots$, where t_i is an admissible sequence connecting s_i and s_{i+2} for every even i, s_i and s_{i-2} for every odd i, t_1 connects s_1 and s_0, and s_i' connects s_i with itself.
2. the periodic points of f are dense.
   Denoting the set of periodic points of f by $Per(f)$, i.e. $y \in Per(f) \Leftrightarrow \exists q, f^q(y) = y$, we have to prove that $\forall x \in \Sigma_1, \forall \varepsilon > 0, \exists y \in B_\varepsilon(x) \cap Per(f)$.
   We work with the metric $d(x,y) = \sum_{i=-\infty}^{\infty} \frac{1}{4^{|i|}} |x_i - y_i|$ defined on $\mathcal{C}$.
   Let us take a bi-infinite sequence x of $\mathcal{C}$. We have to construct a y belonging to $B_\varepsilon(x)$ and to $Per(f)$. It is sufficient that y matches x on a central part (around index 0) of length $2l + 1$ to guarantee $y \in B_\varepsilon(x)$.
   If y is the bi-infinite repetition of $x_{-l} \cdots x_{-1} x_0 x_1 \cdots x_l$, it is $(2l + 1)$-periodic but also $m(2l + 1)$-periodic. Thus, we have $y = \rho^{m(2l+1)}(y) = f^{n(2l+1)}(y)$. Hence y is periodic for f.
   To treat the subshift case, just add a sequence $c_1 \cdots c_K$ in y, where K is less than the smallest exponent k such that the transition matrix of the subshift is irreducible.

Remark 8.15. Similar results have been proved in [47, 213].

Is their long-term behavior very regular or chaotic (see Fig. 8.2 and the formal definition of shifts in §8.4)? Why does this paradox exist?

The above proof of chaotic behavior is based on a specific metric assigning weights to the cells of the cellular automaton, which strongly influences the consideration made, whereas we do not have this weighting in mind when we observe the successive configurations generated. More precisely, the metric space is based on the astronomer's metric (faraway objects are negligible, see Def. 2.53), which does not seem to be the adequate choice. Thus, we have to find a technical approach closer to our own observation and interpretation.

We have proposed to use attraction as technical tool but, since we work with bi-infinite lattices of cells, we considered transfinite iterations and attraction. We need a second tool, namely shifted Hamming distance, which is also very close to the intuition, without the need of transfinite iterations.

Below, we present three structural organizations of CA classes that allow us to partition them into categories on which mathematical as well as intuitive criteria agree.

8.5.2 Linear Periodicity Hierarchy

Although we have the following inclusions:

$$\mathcal{N} \subset \mathcal{F} \subset \mathcal{P},$$

it is difficult to compare the first classes with type $\mathcal{S}$ and type $\mathcal{A}$.

However, if we try to see all classes w.r.t. transfinite iterations, we can see a hierarchy of periodic systems. From type $\mathcal{N}$ to type $\mathcal{A}$, the period grows from one to an ordinal "close" to the cardinality of the configuration space, and the resulting attractor grows from a homogeneous fixed configuration to the whole configuration space. We quote "close" because, for example, we allow to say that $\frac{\aleph_1}{2}$ is closer to $\aleph_1$ than to $\aleph_0$.

In each class, we have an expression like $f^\alpha(\mathcal{C}) = \beta_n$ or $\mathcal{C} \rightsquigarrow^f_\alpha \beta_n$ where $\alpha \in \{\omega, \eta_{\mathbb{P}(\mathcal{C})}\}$, $\eta_{\mathbb{P}(\mathcal{C})}$ is the least transfinite number greater than the number of states of the lattice $\mathbb{P}(\mathcal{C})$, $\beta_n \in \{\text{point}, \text{set}\}$, and β_n contains n-periodic points of f. Summarizing these data according to our classification yields Table 8.1.

Table 8.1. Linear periodicity hierarchy: properties of the classes

	α	β_n	n
$\mathcal{N}$	ω	point	1
$\mathcal{F}$	ω	{ point }	1
$\mathcal{P}$	ω	{ set }	$\in \mathbb{N}$
$\mathcal{S}$	η	{ point }	1
$\mathcal{A}$	η	$\mathcal{C}$	$\in \mathbb{O}$

Pairs (α, β_n) are ordered as follows:

$$(\alpha, \beta) \ll (\alpha', \beta') \text{ iff } \begin{cases} \alpha \ll \alpha' \\ \text{or} \quad \alpha \neq \alpha' \text{ and } \beta \ll \beta' \end{cases}$$

where, by definition, $\omega \ll \eta$ and point $\ll$ {point} $\ll$ {set} $\ll \mathcal{C}$. Hence, we have the linear hierarchy presented in Table 8.2.

Table 8.2. Linear periodicity hierarchy

$$\boxed{\;\mathcal{N} \ll \mathcal{F} \ll \mathcal{P} \ll \mathcal{S} \ll \mathcal{A}\;}$$

8.5.3 Periodicity Clustering

In this second organization, we introduce two criteria:

- individual (in)dependence to initial conditions
- global (trans)finite attractor.

They permit to build a classification table of our different types of behaviors (see Table 8.3).

Table 8.3. Periodicity clustering

Periodicity	Dep. to I.C.	Indep. to I.C.
Finite	$\mathcal{F}$, $\mathcal{P}$	$\mathcal{N}$
Transfinite	$\mathcal{A}$	$\mathcal{S}$

Here, it is interesting to remove the distinction finite/transfinite, which is natural if we accept transfinite iterations as an iteration scheme.

Actually, $\mathcal{F}$ is a part of $\mathcal{P}$. We could divide $\mathcal{P}$ into a countable infinity of disjoint subsets: $\mathcal{P} = \cup_{n \in \mathbb{N}} \mathcal{P}_n$, where $\mathcal{P}_n$ is the set of $n-$periodic CA, n being the maximal period. We have thus the equality $\mathcal{F} = \mathcal{P}_1$. From one to $\eta_{\mathbb{P}(\mathcal{C})}$, there is a whole family of periodic CA. Type-$\mathcal{A}$ CA are the limit of periodic behaviors: $\mathcal{A} \approx \mathcal{P}_{\#\mathcal{C}}$, which gives them a typical aperiodicity.

We have also a class grouping $\mathcal{N}$ behaviors with $\mathcal{S}$ behaviors. From the point of view of this organization, subshift behaviors are very simple.

A last remark about subshift behaviors is the following. We claim that $\mathcal{S}$ behaviors are independent from initial conditions. In fact, it is true on a subspace Σ_1 of the whole configuration space. On its complement, no general statement can be made: the behavior can be aperiodic or periodic, or convergent to Σ_1. Again, this is opposed to the idea that shifts are chaotic.

8.5.4 Organization w.r.t. Shifted Hamming Distance

Let us now come back to the third class of behaviors mentioned in §8.4 (intermediate). These rules are chaotic, as we proved it (see Theorem 8.14). However, when we observe their long-term behavior, what we see is a perfect regularity.

With the help of the tool previously introduced, we can classify our different behaviors in a very simple way. Indeed, thanks to the shifted Hamming distance, we are able to show that subshifts behaviors are simple. The intuition is the following. Let us consider a very simple shift behavior, say the classical left shift. Although this shift can be proved chaotic under some assumptions using an astronomer's metric, the patterns observed when the shift evolves make us think that the system is simple: there is just a shift to the left at each step. If we take the sequence obtained after a few steps, it very much resembles the initial one: we just have to shift it to the right to make it equivalent to the former again. This is the idea behind the shifted Hamming distance: a Hamming distance forgetting the shifting motions.

If the system is n-periodic or if it has a generalized subshift behavior including a n-periodicity, they both appear very simple through the shifted Hamming distance: for any x in the orbit of an initial condition, after the transient,

$$H^\rho(x, f^n(x)) = 0.$$

The dual behavior is aperiodic. This gives us another characterization for complexity: there is no x in the configuration space $\mathcal{C}$ for which the previous condition applies and thus, for every x, and every n, we have:

$$H^\rho(x, f^n(x)) \neq 0.$$

It might be interesting to study how far this distance is from zero. Some systems have perhaps a bounded SHD whereas other ones could reach 1. This could help in a deeper analysis of type-$\mathcal{A}$ CA.

We summarize this last organization in Table 8.4: under astronomer's metrics, subshifts can be considered as chaotic, whereas under the SHD, subshifts are again very simple, just as periodic behaviors.

Table 8.4. Organization w.r.t. shifted Hamming distance

Null SHD	Positive SHD
$\mathcal{N} \cup \mathcal{F} \cup \mathcal{P} \cup \mathcal{S}$	$\mathcal{A}$

This organization shows that the SHD can be an interesting tool to study CA behaviors, especially to refine subclasses of type-$\mathcal{A}$ CA. Actually, it should be very useful to investigate the quotient space $\mathcal{C}/_{\equiv_\rho}$. What kind of behaviors does this space allow? What is complexity in this case? This is left for future work.

8.5.5 Dynamical Complexity in CA

The question of complexity is not easy to address, even for elementary CA; the many existing classifications evidence this (see §8.10.3). Moreover, sometimes, using classical tools, the resulting classification is not intuitive.

That was our main motivation to introduce a formal classification of CA behaviors corresponding to phenomenological considerations. We needed new tools, namely transfinite attraction and shifted Hamming distance; the former allowing a classification in terms of transient length before convergence to cycles, the latter showing that (periodic) subshifts are simple. Based on this, the following complexity hierarchy can be defined.

Definition 8.16 (Complexity). *On the set $\{ \mathcal{N}, \mathcal{F}, \mathcal{P}, \mathcal{S}, \mathcal{A} \}$, the following ordering is established: $\mathcal{N} < \mathcal{F} < \mathcal{P} < \mathcal{S} < \mathcal{A}$. The complexity $\Upsilon(g)$ of a CA based on the local rule g is the place of its long-term behavior in the defined ordering.*

8.6 Conjectures in CA Composition

Each CA rule can be represented by a vector of eight binary components, corresponding to all possible neighborhood configurations (see §8.2.1). Using the componentwise logical disjunction, it is possible to generate all elementary rules from the basis $\{1, 2, 4, 8, 16, 32, 64, 128\}$. In [102, 101], the dynamics of these particular CA rules was further studied separately and under composition. From the systematic compositions of basis rules, conjectures were proposed that very much resembles the results presented in Chap(s). 6 and 7 union invariants. The aim of this section is to motivate a deeper study of disjunction by comparison to union.

Basis Rules. The basis contains elements from each of the first four classes defined in §8.4: using [49]'s notations, it gives

$$(1 \in p, 2 = s^+, 4 \in f, 8 = n^+, 16 = s^-, 32 \in n, 64 = n^-, 128 \in n)$$

and, using our standard notation,

$$(8, 32, 64, 128 \in \mathcal{N}, 4 \in \mathcal{F}, 1 \in \mathcal{P}, 2, 16 \in \mathcal{S}).$$

Remark 8.17. Rules denoted by n^+, n^- and s^+, s^- are symmetric: their local transition functions are equal under left-to-right transformation, for instance, $g_{n^+}(a, b, c) = g_{n^-}(c, b, a)$; this property is important when observing the composition of rules.

Table 8.5. Local rule tables of CA 2, 16 and 18

Rule	000	001	010	011	100	101	110	111
2	0	1	0	0	0	0	0	0
16	0	0	0	0	1	0	0	0
18	0	1	0	0	1	0	0	0

Disjunction. This composition is easy to compute: Table 8.5 shows the disjunction of rules 2 and 16, the result being rule 18.

Local and global disjunctions are equivalent: let g_1 and g_2 be two local rules, and x be a configuration of $\mathcal{C}$, then

$$(\otimes_R g_1)(x) \vee (\otimes_R g_2)(x) = (\otimes_R (g_1 \vee g_2))(x).$$

Notice that this will allow us to merge G and G' as well as $\star$ and $\diamond'$ in the abstract expressions of §8.1.

Complexity and Disjunction. Let us now present all binary disjunctions of basis rules in terms of the classification given in §8.4. We intend to derive information about the dynamics of CA considering the dynamics of the single basis rules in the composition. The abstract expression presenting CA composition in §8.1 becomes

$$\Upsilon(\otimes_R(\vee_i g_i)) = \Upsilon(\vee_i(\otimes_R g_i)) = \diamond_i \Upsilon(\otimes_R g_i),$$

where $\diamond$ is obtained by exhaustive simulation of all possible binary disjunctions (see Table 8.6, where symmetric cases are omitted for clarity), and Υ refers to Def. 8.16.

Table 8.6. Local ∨-composition

		8 n^+ $\mathcal{N}$	32 n $\mathcal{N}$	64 n^- $\mathcal{N}$	128 n' $\mathcal{N}$	4 f $\mathcal{F}$	1 p $\mathcal{P}$	2 s^+ $\mathcal{S}$	16 s^- $\mathcal{S}$
8	$\mathcal{N}$	-	$\mathcal{N}$	$\mathcal{F}$	$\mathcal{N}$	$\mathcal{F}$	$\mathcal{S}$	$\mathcal{S}$	$\mathcal{S}$
32	$\mathcal{N}$	-	-	$\mathcal{N}$	$\mathcal{N}$	$\mathcal{F}$	$\mathcal{P}$	$\mathcal{S}$	$\mathcal{S}$
64	$\mathcal{N}$	-	-	-	$\mathcal{N}$	$\mathcal{J}$	$\mathcal{S}$	$\mathcal{S}$	$\mathcal{S}$
128	$\mathcal{N}$	-	-	-	-	$\mathcal{F}$	$\mathcal{A}$	$\mathcal{S}$	$\mathcal{S}$
4	$\mathcal{F}$	-	-	-	-	-	$\mathcal{P}$	$\mathcal{S}$	$\mathcal{S}$
1	$\mathcal{P}$	-	-	-	-	-	-	$\mathcal{S}$	$\mathcal{S}$
2	$\mathcal{S}$	-	-	-	-	-	-	-	$\mathcal{A}$
16	$\mathcal{S}$	-	-	-	-	-	-	-	-

Conjectures. Looking at Table 8.6, we see that when two symmetric rules are composed together, complexity grows:

$$\underbrace{\otimes_R 8}_{n^+ \in \mathcal{N}} \vee \underbrace{\otimes_R 64}_{n^- \in \mathcal{N}} = \underbrace{\otimes_R (8 \vee 64)}_{72 \in \mathcal{F}}$$

$$\underbrace{\otimes_R 2}_{s^+ \in \mathcal{S}} \vee \underbrace{\otimes_R 16}_{s^- \in \mathcal{S}} = \underbrace{\otimes_R (2 \vee 16)}_{18 \in \mathcal{A}}.$$

Apart from this, type $\mathcal{F}$ is obtained from type $\mathcal{F}$, and periodicity entails type-$\mathcal{P}$, type-$\mathcal{S}$, and type-$\mathcal{A}$ behaviors. Two important complexity laws were conjectured in [49, 101].

Conjecture 8.18 (Law of complexity conservation). *In the local disjunctive composition of two basis rules, the complexity never decreases:*

$$\forall x, y \in \{\mathcal{N}, \mathcal{F}, \mathcal{P}, \mathcal{S}\}, \Upsilon(x \vee y) \geq \max(\Upsilon(x), \Upsilon(y)).$$

Conjecture 8.19 (Law of complexity increase). *The complexity of the local disjunctive composition of two basis rules increases (e.g., $\Upsilon(x \vee y) > \max(\Upsilon(x), \Upsilon(y))$) iff one of the following conditions holds:*

$- x = n^+$, $y = n^-$;	*(symmetric null rules)*
$- x = s^+$, $y = s^-$;	*(symmetric shifting rules)*
$- x \in \mathcal{N} \backslash \{32\}$, $y \in \mathcal{P}$.	*(periodicity)*

We see that two particular cases appear to increase complexity: concurrent symmetry and periodicity. In the rest of this section, we concentrate on the first case, which seems to behave as prescribed by union-invariant theorems.

8.7 Complexity by Composition of Shifts

Among the experimental facts described in the previous section, let us pay attention to Conj. 8.19, and to its second statement in particular: disjunction of shifts leads to complexity.

Here, we proceed to a naive comparison of this conjecture to the compositional analysis of the Cantor relation or any other complex system obtained by union of elementary systems attracting the space to different fixpoint invariants (see e.g. §7.3).

8.7.1 Rules 2 and 16

The rule tables of 2 and 16 are represented in Table 8.5, together with their disjunction, viz. rule 18. Rule 2 gives 1 iff the local configuration is $(0, 0, 1)$, 0 in the other cases; rule 16 gives a 1 in the symmetric configuration: 1

iff the local configuration reads $(1,0,0)$, 0 otherwise. We then compute the disjunction of rules 2 and 16, which gives rule 18: 1 iff the local configuration is $(1,0,0)$ or $(0,0,1)$, and 0 otherwise.

The behavior of rules 2 and 16 is very simple (type $\mathcal{S}$): configurations are shifted along the lattice, to the left or to the right. The behavior of rule 18 is less simple, we classify it as a complex, type-$\mathcal{A}$ CA. Examples are represented in Fig. 8.4.

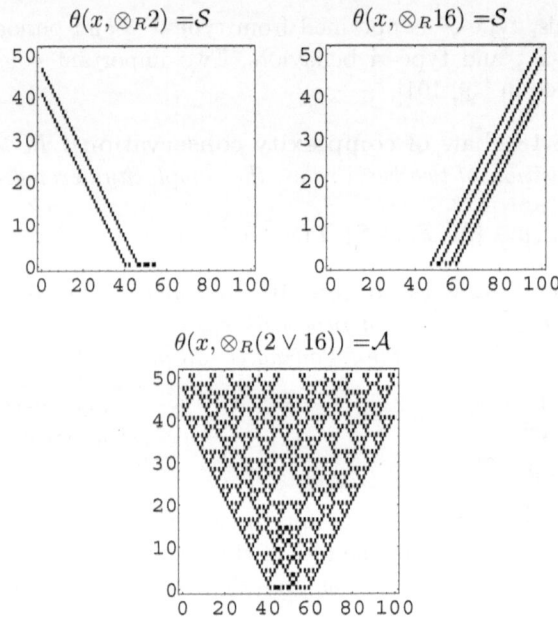

Fig. 8.4. Evolution of CA rules 2 (left) and 16 (right) from a random initial condition x show simple shifting behaviors, their disjunction 18 (bottom) shows a complex type-$\mathcal{A}$ dynamics

8.7.2 Cantor Relation

Let us now briefly summarize the compositional analysis of the Cantor relation f (see also §7.3), in order to emphasize how close its dynamics is to the disjunctive composition (see Fig. 8.5).

The relation is defined by union composition:

$$f_1 = \frac{x}{3} \tag{Int}$$

$$f_2 = \frac{x+2}{3} \tag{Int}$$

$$f = f_1 \cup_F f_2. \tag{Int}$$

Proposition 8.20. *Cantor relation* $f = f_1 \cup_F f_2$ *has a complex dynamics determined by a Cantor-set invariant.*

Proof. The behavior of f_1 on $[0, 1]$ is very simple: every point is asymptotically attracted to the fixed point 0. Function f_2 has also a unique attracting fixed point on $[0, 1]$, which is 1. These fixpoints are different. Moreover, f_1 and f_2 are both contracting, with a coefficient $\gamma(f_1) = \gamma(f_2) = \frac{1}{3}$, and they verify $f_1^{-1}([0, 1]) = f_2^{-1}([0, 1]) = [0, 1]$. All assumptions of Theorem 6.28 are verified, which concludes the proof.

The behavior of the union is much richer than its individual components: it is has a Cantor-set invariant attracting the whole space.

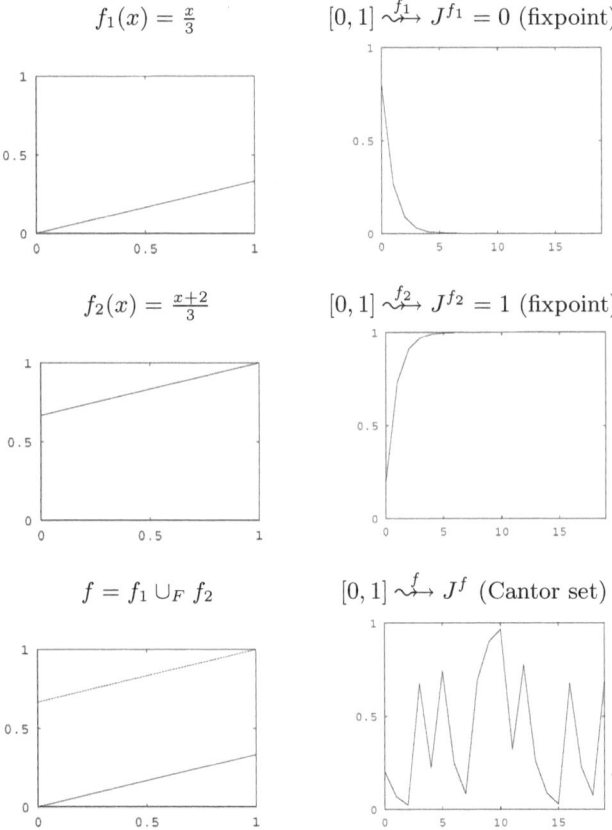

Fig. 8.5. Compositional analysis of the Cantor relation: f_1 and f_2 show simple dynamics (attraction to their respective fixpoints); f shows a complex (direct and reverse) dynamics, i.e. chaotic on the Cantor set invariant.

8.7.3 Comparison

Rules 2 and 16 behave in a very simple way: every configuration is shifted to the left or to the right. Everything seems to be attracted to the same point: infinity. Regarding f_1 and f_2, every point of $[0,1]$ is continuously attracted to 0 or 1.

Rule 18 is obtained by disjunction of rules 2 and 16. Globally, one can consider that at some places of a configuration, the behavior of 2 is executed, whereas at other places the behavior of 16 is executed. The same thing appears when executing the union of two systems. An important difference is that the choice in the first case is directed by the opportunity to activate a transition, while in the second case the choice is directed by an oracle (pure randomness in the future, but recall the system is deterministic backward and yet complex).

Finally, when rules 2 and 16 on one hand, or f_1 and f_2 on the other hand, are composed together such that both components can be applied on the configuration at the same time, complexity arises. In the first case, this is only an experimental conjecture (Conj. 8.19) which is sort of evidenced by simulation, whereas in the second case, there is a theorem to prove the observation (Theorem 6.28 and Cor. 6.29).

8.7.4 A More Precise Conjecture

Considering the first experimental results and rephrasing the essence of union-invariant theorems in the context of CA, Conj. 8.19 can be refined as follows.

Conjecture 8.21 (Complex dynamics by $\vee$-composition − 1). *Rules 2 and 16 have compatible dynamics (contracting in the future), their disjunction is globally contracting in the future, and their invariants are different fixpoints. Then $\otimes_R(2 \vee 16)$ has a rich invariant and a complex dynamics.*

Remark 8.22. This complex dynamics is not necessarily related to chaos as defined in §5.2 because of the problem mentioned in §8.5 regarding type-$\mathcal{S}$ CA.

8.8 Qualitative Analysis and Complexity Measures

Now we have more precisely stated that the disjunctive composition of shifting CA leads to complexity (Conj. 8.21), the questions are:

− Does rule 18 really behave like the Cantor relation?
− How to prove that the resulting dynamics is complex?

Qualitative Analysis. The problem is that we do not have any theorem to treat this case of local composition. In other words:

– Does rule 18 behave like a global union?
– In mathematical terms: $\Upsilon(\otimes_R 18) = \Upsilon((\otimes_R 2) \cup (\otimes_R 16))$?

Probably not! In fact, rule 18 certainly entails more complexity than a global union because it acts locally. We thus investigate the intermediate case between a global union and this local disjunction: a local union.

The reason why we decide to study this intermediate case amounts to the fact that these three versions (local disjunction, local union, global union) can be seen as particular cases of a more general composition type that very much resembles probabilistic CA.

Generally, we can define a *probabilistic CA* as follows: at each step t, each cell i applies a local rule g depending on t and i. This local rule is a probabilistic choice between (at least) two possibilities: g_1 with probability p and g_2 with probability $1-p$. The model can be reduced to four specific cases, according to when and where the probabilistic choice is made. This leads to Table 8.7, where we mention an (approximately) equivalent nonprobabilistic model.

Table 8.7. Different models of probabilistic CA

place of choice	time of choice	choice	$\approx$ model
local	each step	g_i^t	local union
local	once	g_i	interleaved [101]
global	each step	g^t	global union
global	once	g	classical CA

Complexity Measures. We have to characterize the complexity of different systems. Generally, we base our intuitive understanding of complexity on the notions of invariance and attraction. Here, we focus on the richness of the attractor, but we introduce other tools from the experimental part of CA studies:

– Boolean derivative and weight;
– generalized mean-field theory;
– entropy.

Boolean Derivative and Weight. In the following, we apply the notion of Boolean derivative to CA [312]. Given a Boolean function f, its *Boolean derivative* is defined by

$$\frac{\partial f}{\partial x_j} = f(\cdots \widehat{x_j} \cdots) \oplus f(\cdots x_j \cdots)$$

$$\dot{f} = (\frac{\partial f}{\partial x_{j-1}}, \frac{\partial f}{\partial x_j}, \frac{\partial f}{\partial x_{j+1}})$$

where $\oplus$ is the exclusive disjunction, and $\widehat{}$ stands for a logical negation.

Since every elementary rule can be expressed as an exclusive disjunction of basic rules, the derivative is easy to compute: R_i being basic rules, c_i being binary coefficients,

$$f = \oplus_i c_i R_i \Rightarrow \dot{f} = \oplus_i c_i \dot{R}_i.$$

The *weight* of a function f is the average number of 1's in the partial derivatives of f. It is related to the sensitivity to initial conditions of the global function. The bigger is the weight, the more sensitive the function is.

Generalized Mean-Field Theory. To evaluate the disorder induced by a rule, several methods are conceivable. In [137, 136], the authors present the *generalized mean-field theory*.

- The 0th order corresponds to Langton's λ parameter [195, 133]: it takes the proportion of 1's in the image of all triples of the rule.
- The 1st order stands for a mean-field approximation of the 0th order.
- The 2nd order is an extension of the first idea to larger neighborhoods: it involves spatial correlations.
- As further orders converge to an invariant measure, they mimic the behavior of cellular automata with more and more fidelity.

Entropy. A notion of entropy is also useful to characterize the complexity or disorder induced by some rules (this kind of notion has been extensively studied in [203, 50]). If $(p_i)_{i \in I}$ is a probability distribution corresponding to a set of possible events I, the *entropy* of this distribution is:

$$S = -\sum_{i \in I} p_i \log(p_i).$$

In this case, probabilities could be Markov approximations of CA rules as given by the generalized mean-field theory.

8.9 Compositional Analysis of Complex CA

In this section, we systematically analyze the local disjunction, the local union, and the global union using the complexity measures introduced in §8.8. This allows us to reinforce Conj. 8.21.

8.9.1 Local Disjunction, Local Union, and Global Union

Because of the model itself, it is obvious that the complexity of the global union is smaller than the complexity of the local union:

$$\Upsilon((\otimes_R 2) \cup (\otimes_R 16)) < \Upsilon(\otimes_R (2 \cup 16)).$$

What we would like to show goes in the same direction:

$$\Upsilon(\text{global union}) < \Upsilon(\text{local union}) < \Upsilon(\text{local disjunction}).$$

To compare these different cases, we use the complexity measures introduced above. For each rule, we have:

- a table with, in each column,
    - a typical neighborhood configuration;
    - the local image the rule will produce after one iteration;
    - a label given to the neighborhood;
    - the possible labels appearing when the label itself is embedded in a bigger neighborhood of radius 2 (instead of radius 1), after one iteration;
    - the probability to find the label after one iteration, starting from a neighborhood of radius 2;
- the labels that are never reached in an evolution;
- the Boolean derivative of the rule, and its corresponding weight;
- the proportion of 1's in the image of the rule (Langton's λ parameter);
- the entropy of the rule, based on the probabilities given above.

In Tables 8.8 to 8.11, we detail the analysis of four different rules, namely rules 2, 16, their local disjunction and their local union.

Table 8.8. Rule 2, $\otimes_R 2$

	000	001	010	011	100	101	110	111
image	0	1	0	0	0	0	0	0
label	a	b	c	d	e	f	g	h
next	a,b	c	a,e	a,e	a	a	a	a
proba	$\frac{20}{32}$	$\frac{4}{32}$	$\frac{4}{32}$	0	$\frac{4}{32}$	0	0	0

Table 8.9. Rule 16, $\otimes_R 16$

	000	001	010	011	100	101	110	111
image	0	0	0	0	1	0	0	0
label	a	b	c	d	e	f	g	h
next	a,e	a,e	a,b	a	c	a	a,b	a
proba	$\frac{20}{32}$	$\frac{4}{32}$	$\frac{4}{32}$	0	$\frac{4}{32}$	0	0	0

Table 8.10. Rule 18, $\otimes_R(2 \vee 16)$

	000	001	010	011	100	101	110	111
image	0	1	0	0	1	0	0	0
label	a	b	c	d	e	f	g	h
next	a,b,e,f	c,g	a,b,e,f	a,e	c,d	a	a,b	a
proba	$\frac{14}{32}$	$\frac{4}{32}$	$\frac{4}{32}$	$\frac{2}{32}$	$\frac{4}{32}$	$\frac{2}{32}$	$\frac{2}{32}$	0

Table 8.11. Local union, $\otimes_R(2 \cup 16)$

	000	001	010	011	100	101	110	111
image	0	$\{0,1\}$	0	0	$\{0,1\}$	0	0	0
label	a	b	c	d	e	f	g	h
next	a,b e,f	a,c e,g	a,b e,f	a,e	a,b c,d	a	a,b	a
proba	$\frac{21.5}{32}$	$\frac{3}{32}$	$\frac{3}{32}$	$\frac{0.5}{32}$	$\frac{3}{32}$	$\frac{0.5}{32}$	$\frac{0.5}{32}$	0

8.9.2 Comparison and Summary of Results

Basing our comparison on the tables presented above, we get the following results (see Table 8.12):

Table 8.12. Summarized compositional analysis

Local rule	2	16	$2 \vee 16$	$2 \cup 16$
Unreached	d,f,g,h	d,f,g,h	h	h
Derivative $\dot{R}(x_1, x_2, x_3)$	$\dot{R}_2$ $(\widehat{x_2 x_3}, \widehat{x_1 x_3},$ $\widehat{x_1 x_2})$	$\dot{R}_{16}$ $(\widehat{x_2 x_3}, x_1 \widehat{x_3},$ $x_1 \widehat{x_2})$	$\dot{R}_2 \oplus \dot{R}_{16}$ $(\widehat{x_2}, x_1 \widehat{x_3} \oplus \widehat{x_1} x_3,$ $\widehat{x_2})$	$-$ $-$
Weight Average	$(2,2,2)$ 2	$(2,2,2)$ 2	$(4,4,4)$ 4	$-$ $-$
Prop. 1's	$\frac{1}{8}$	$\frac{1}{8}$	$\frac{1}{4}$	$\frac{1}{8}$
Entropy	1.07354	1.07354	1.66132	1.12789

– proportion of residual 1's

$$0 < P_2 = P_{16} = P_{2 \cup_F 16} < P_{2 \oplus 16} < P_{orig} = \frac{1}{2};$$

– entropy

$$0 < S_2 = S_{16} < S_{2 \cup_F 16} < S_{2 \oplus 16} < S_{orig} = 2.07944.$$

Thus, the local composition $2 \cup 16$ induces more order than $2 \vee 16$, which seems to be a stronger composition than the first one. Hence, we have the second part of the inequality presented above, i.e.

$$\Upsilon(\text{local union}) < \Upsilon(\text{local disjunction}).$$

Proposition 8.23 (Complex dynamics by $\vee$-composition $-$ 2). *The disjunction of rules 2 and 16 entails a complex behavior:*

$$\Upsilon(\otimes_R(2 \vee 16)) > \Upsilon(\otimes_R(2 \cup 16)) > \Upsilon((\otimes_R 2) \cup (\otimes_R 16)) = \quad complex\ behavior.$$

Proof. Using Theorem 6.28, we establish the complex behavior of the global union; evidently, the local union is more complex than its global counterpart; with the help of our complexity measures, we state that the local disjunction introduces more disorder than the local union; this permits to conclude that this local disjunction has a globally complex behavior.

8.10 Discussion

In this section, we first summarize the chapter, and draw some partial conclusions. Then, we state unsolved problems and open questions. Finally, we present related work in classification, aperiodicity, and composition.

8.10.1 Summary and Partial Conclusion

Classification. Two questions were answered. We showed that subshifts are simple, using a rigorous framework including shifted Hamming distance. We also defined and investigated the class of complex behaviors. This class has to be refined but it constitutes a good starting point to lead us to complexity in CA.

Our goal was to find a classification of elementary cellular automata in which each class is defined by a mathematical expression, which was generally not the case in previous classifications (e.g. [330, 49] and §8.10.3). In particular, we wanted to characterize the most "complex" classes, and we wanted subshifts behaviors to be considered as simple because this better matches qualitative behavioral classifications (although subshifts are proved to be chaotic using classical means: our Theorem 8.14 generalizes results of [41]).

We refined a given classification [330, 49] and we added tools to go deeper [102]: transfinite attraction and shifted Hamming distance. With these tools we proposed a formal characterization of each class and we saw that spatiotemporally chaotic or aperiodic systems can be considered as "periodic" systems with huge periods.

Composition. We carried out a "compositional experimental analysis" combining experimental complexity measures with theoretical compositional results. This led to interesting conjectures on the emergence of type-$\mathcal{A}$ complex behaviors by composition of type-$\mathcal{S}$ simple behaviors: Conj(s). 8.21 and 8.23, refining Conj. 8.19 of [49].

What is the interest of the previous developments? Establishing the complexity of the local disjunction is not a breakthrough because it is already clear visually, under simulation. But, we analyzed it by composition with the help of common experimental tools in the field of cellular automata, viz. complexity measures.

The result is: merging two very simple compatible behaviors attracting the space to different parts of it entails the emergence of complexity since, as in all previous cases (see also §5.7.3), the composition realizes an opportunistic or oracle choice between different components.

An increase in complexity can come from the explosion of dimensions of the state space, or from the mixing generated by the neighborhood relation R and the local transition functions g_i's. In fact, the situation is not totally black or white. Complexity can arise even without mixing, just by the union or another operator. Simplicity can also arise from the connected product. All depends on the way systems are composed, together with some important properties: the attraction to different regions of the global state space. In this case, complexity can be created from very simple systems composed together.

8.10.2 Open Questions

Classification. Of course, there are unsolved questions: our classification is undecidable in general, unless used for very simple systems; we did not prove the completeness of our classification: gaps could exist, i.e. behaviors not included in any class.

Another important aspect to investigate is whether there is a link between our definition of aperiodic systems, and *intuitions* underlying classical definitions of chaotic systems [88, 326]. Are they equivalent, contradictory, or complementary?

Finally, we should study the quotient of the configuration space by the shifted Hamming distance whose equivalence classes are shift-invariant configurations because it could help us to investigate complex behaviors in a simpler way, by decreasing their global complexity. We could also refine the structural organization of CA classes of §8.5 using the pseudo-metric defined in [51, 52].

Composition. In order to analyze the behavior of disjunction, we added complexity measures to our theoretical compositional framework. These complementary notions permit to describe or discover complex behaviors.

We could make use of other complexity measures to refine our results (e.g. higher orders of generalized mean-field theory, variants of entropy), and elaborate stronger techniques for the compositional analysis of connected products in particular. Then, in addition to union (and disjunction), other composition operators could be applied to CA and analyzed in the light of complexity measures, too. Partial experiments have already been conducted in [101, 105] on sequential and disjunctive composition, that confirm the power of mixing theoretical compositional results with complexity measures. However, the exact influence of the elementary periodic rule p in the composition is still unclear.

Finally, although we have considered homogenenous CA, a very interesting open problem is to use the compositional approach to analyze heterogeneous, or *hybrid*, models (i.e., using possibly different local functions at different cells of the automata) [54, 55]. Some research in this direction has already started [101].

8.10.3 Classification: State-of-the-Art

Classification is one of the central themes in the theory of CA. This motivates this rather long section stating important earlier classification schemes.

Several authors have proposed different classifications, starting with Wolfram in 1983. Although we call this section "state-of-the-art", we mainly present classification schemes close to Wolfram's one. Other schemes are mentioned in the papers cited here.

Attraction-Based Classification. In [330], the first classification of CA appears, grouping together systems having the same long-term behavior: class I, evolution to homogeneous state; class II, evolution to separated simple states or periodic structures; class III, chaotic patterns; class IV, complex localized structure, sometimes long-lived. The main problem of this classification is that it is only qualitatively defined. The fourth class is related to universal computational devices. Wolfram considers only symmetric rules, the local rule of which is such that $g(a, b, c) = g(c, b, a)$. When asymmetry appears, we get subshift behaviors for example.

In [174], the author relates the previous classification to two properties: the number of attractors, related to storage capacities when the systems are considered as associative-memory devices; the period size, related to information dynamics. This gives a more precise description of each class (see Table 8.13).

In [200], the authors give a classification looking like Wolfram's one but not completely comparable to it: class 1, null rules, leading to homogeneous configurations (equivalent to Wolfram's class I); class 2, evolution to fixed-point configurations (class II, partly); class 3, evolution to periodic configurations (class II); class 4, locally chaotic rules (class II); class 5, globally chaotic rules (class III); class 6, complex behaviors (class IV). They also study inter, and intra-class probabilities and give a mean-field description of their classification.

Table 8.13. Kaneko's attraction-based classification

	Small number of attractors	Large number of attractors
Short periods	no information creation small storage class I	no information creation large storage class II
Long periods	creation of information small storage class III	creation of information large storage class IV

In [49], a variation of Wolfram's classification is presented, adding classes behaving like subshifts, due to the consideration of asymmetric rules. Here, the authors take bi-infinite configurations without any restriction: class n, evolution to quiescent configurations (Wolfram's class I); class f, evolution to fixed-point configurations (class II, partly); class p, evolution to periodic configurations (class II); class s, simple subshift behaviors (class II); class s', complex subshift behaviors (class II); class c, "chaotic" behavior (classes III and IV). Classes s and s' are considered separately from simple rules because, when studied on bi-infinite configurations, they generate chaos (as proved in §8.5, see Theorem 8.14), whereas they are seen as very simple rules when evolving from finite configurations. This classification was our starting point. We will come back to this below.

In [42], the authors propose a new classification of CA, based on the observation of finite initial conditions in bi-infinite configurations. The tool presented is a measure of pattern growth. The classification goes as follows: class C_1, patterns vanish; class C_2, pattern length stays finite (fixed or periodic finite size); class C_3, pattern length grows to infinity. We have the following relation between classes: $C_1 \subset C_2 = \overline{C_3}$. An important advantage of this classification over other ones is that it is decidable. For each rule, it is possible to determine a priori whether it belongs to class C_1, C_2, or C_3. Since this classification is decidable, let us show some links with ours: $\mathcal{N} = C_1$, $\mathcal{F}$, $\mathcal{P}$, $\mathcal{S}$ are in C_2, and $\mathcal{A} \cap C_3 \neq \emptyset$. This classification is decidable because it is based on a subset of the whole configuration space, namely finite initial conditions.

Finally, in [171], a comprehensive classification is given, based on three kinds of properties, namely: formal languages (3 classes), equicontinuity (4 classes) and attraction (5 classes). To our knowledge, this is the finest classification up to now. Sixty different classes of behaviors are analyzed. Examples and counter-examples are exposed. Only three classes are not completely characterized in the sense that it has not yet been proved whether they are empty or not.

Decidability Results. There are many papers related to decidability results. Since this aspect is somewhat outside our topic, we just mention two papers, wherein the interested reader can find a good initial list of references.

In [74] the authors give a hierarchy of CA starting with finite configurations: class I, CA converging to fixed-points in finite time; class II, CA converging to periodic configurations in finite time; class III, CA for which it is decidable to know whether a configuration occurs in the orbit of another; class IV: all CA. Each class contains the previous one(s). The authors emphasize (un)decidability results

In [301], the author concentrates on circular CA and uses the same definitions as in [74].

Probabilistic Approach. Many authors make use of probabilities, Markov chains, and statistical mechanics in the field of CA, which give very interesting global results. Since this approach is orthogonal to ours, we do not mention many references. We refer the interested reader to the following two papers, in which many other references can be found.

In [137, 136], the authors make use of mean-field theory to characterize and classify CA. In [134], the author studies the action of CA on n-step Markov measure.

Summary. Let us summarize what the problems are in general and compare our approach with earlier work. Two problems appear when classification is studied: it is difficult to give a formal definition of each class of CA (in particular, spatiotemporal chaos is not precisely defined in this context); these definitions are often based on undecidable properties. Our classification is strongly influenced by [49]. We put together classes s and s' and give a generalized version of subshift behaviors. We also generalize class c into a class of aperiodic behaviors (see [161] below). We present tools allowing to give a precise formal definition of each class. These tools can be related to the characterization appearing in [174]. Our classification is not decidable for all CA rules but only for basic ones. We have partially extended these results to more complex rules, with the help of composition operators.

8.10.4 Aperiodicity in Cellular Automata

The notion of "chaos" is still not well defined in the context of discrete-time discrete-space multi-dimensional dynamical systems such as, for example, cellular automata. Several authors propose ways of defining complex behaviors in CA. This is one of the goals of classification. We have already presented several classification schemes. Other ones emphasize transition phenomena in the space of CA rules, allowing new classifications, too. In these latter ones, statistical measures are often used, together with information theory-like measures (entropy, activity, sensitivity to rule change, etc.). This leads to definitions of complex behaviors, based on certain parameter values. Among others, we refer the interested reader to [63, 125, 201, 257, 332].

In [302], the author presents a classification of chaotic behaviors, based on notions of randomness, complexity measures, computability of initial conditions, and (non)determinism of rules.

Finally, in [161], the author studied aperiodicity of some CA analytically. We take this point of view in our classification scheme because it is easier to define than complex or chaotic rules. However, we do not make use of linearity and injectivity notions presented by Jen. This point of view is interesting because aperiodicity includes complex and chaotic behaviors. Recently, we have extended the analysis to continuous CA [103].

8.10.5 Related Work in Composition

In the literature on cellular automata and related models, lots of papers study complexity. Among these ones, let us just refer to important approaches developed in [125, 330, 63, 177]. Two books offer many contributions and references [132, 113].

Closer to our compositional approach, we find algebraic attempts to characterize the behavior of some CA in [53, 249, 315]. Composition operators are proposed, together with global results obtained by composition of local properties. A comprehensive, recently published book further develops these aspects [317].

Finally, [49] is the starting point of the experimental part of this chapter, giving the same results by simulation as the ones mentioned in the theoretical framework of [290], basically contained in Chap(s). 2 to 7. A common work further led to [102, 101]. More recently, a characterization of dynamical and algebraic properties was proposed for a particular class of shifting rules, namely moss reinforced shifts, based on the framework presented in this chapter [104].

9. Compositional Analysis of Computational Properties

We studied the rich diversity of behaviors of dynamical systems via their invariants and attractors, and we examined how these properties, related to dynamical complexity, are combined when systems are composed together using appropriate composition operators. This was the aim of Chap(s). 5 and 6, illustrated by case studies in Chap(s). 7 and 8.

After this study of dynamical properties of systems, we examine a second important aspect of dynamical systems, namely their computational abilities. We embed classical computational models in a uniform structure of composed dynamical systems and we compare their computational power with two representative classes of systems: cellular automata and continuous functions. The comparison we carry out is based on extrinsic (simulation) and intrinsic (topological, metric and computational) properties. This allows us to propose a hierarchy of computational models that are completely characterized by composition.

The chapter is organized as follows. We start in §9.1 with an introduction motivating the comparison between computational models and dynamical systems; in §9.2, we informally describe how systems can be compared; in §9.3, we embed Turing machines, cellular automata and continuous functions in a generic composed system based on the connected product; in §9.4–§9.6, these families of systems are compared w.r.t. simulation, which leads to a weak hierarchy, using topological and metric properties, allowing a precise characterization of their computational power, and regarding their behavior from (un)computable initial conditions; in §9.7, the results of this chapter are summarized and a hierarchy of systems is proposed; finally, we close the chapter in §9.8 with a discussion.

9.1 Automata as Dynamical Systems

In this section, we introduce and motivate the elaboration of a computation hierarchy of dynamical systems. Starting from qualitative properties of systems such as invariance and attraction (see Chap. 5), we obtain language-theoretic properties that allow us to compare classical computational models like Turing machines with other models like cellular automata and continuous functions.

F. Geurts: Abstract Compositional Analysis of Iterated Relations, LNCS 1426, pp. 217-242, 1998.
© Springer-Verlag Berlin Heidelberg 1998

If one looks at the dynamics itself as an input/output relation, one can wonder

- what kind of attractor a system is able to reach;
- what set of initial states will lead to a given set of final or asymptotic states.

Actually, these questions only rephrase attraction properties:

- given a set of initial conditions I, solve $I \rightsquigarrow ?$;
- given a set of final conditions F, solve $? \rightsquigarrow F$;

but we do not focus anymore on the evolution leading to the attractors; contrarily, things are considered globally.

The questions can still be rephrased differently, asking what kind of relation f, leading to a behavior like $I \overset{f}{\rightsquigarrow} U$ or $V \overset{f}{\rightsquigarrow} F$, we could realize with dynamical systems. This is getting closer to computability theory, where the questions concern the type of functions one can build using a given computational model.

For example, let us remember that *Chomsky's hierarchy* establishes a strict ordering between the computational power of different models including finite automata, pushdown automata, and Turing machines [155]. Of course, the precise meaning of "computational power" has to be defined in order to carry out these comparisons. In this case, one is interested in languages recognized (accepted) or generated by different models. Let us examine some simple examples.

Example 9.1 (Nondeterministic finite automaton). A nondeterministic finite automaton is defined by a set of states Q, an alphabet Σ, a transition relation $\zeta \subseteq (Q \times \Sigma^* \times Q)$, an initial state $q \in Q$, and a set of accepting states $F \subseteq Q$. The automaton starts in state q and is provided with an input sequence $s \in \Sigma^*$. At each step, a transition $(s, q) \rightarrow (s', q')$ occurs if there exists a word $w \in \Sigma^*$ such that $s = ws'$ and $(q, w, q') \in \zeta$. It is possible to rewrite this nondeterministic transition as a choice in the relational iteration of the system f_A working on $\Sigma^* \times Q$:

$$f_A = \{((s, q), (s', q')) \mid \exists w \in \Sigma^*, s = ws' \land (q, w, q') \in \zeta\}.$$

The language defined as the set of inputs leading to an accepting state in a finite amount of steps is said to be "recognized", and is denoted by L_r:

$$L_r \times \{q\} \overset{f_A}{\rightsquigarrow} \Sigma^* \times F.$$

This language is nothing but the backward attractor of $\Sigma^* \times F$. If we reverse the execution of f_A by taking its inverse f_A^{-1}, L_r is the forward attractor.

Example 9.2 (Regular grammar). We consider now regular grammars generating regular languages (other types of grammars can be treated in the same way). Such a formal system is defined by an alphabet V, a set of terminal symbols $\Sigma \subseteq V$, an axiom S and a set of inference rules $R \subseteq V^+ \times V^*$. There is a constraint on R: the only possible rules are either $A \rightarrow wB$ or $A \rightarrow w$, with A and B being nonterminal symbols, and $w \in \Sigma^*$. The initial symbol is S, and all derivations start from it. A derivation $u \rightarrow v$ occurs when there exists a pair of words $x, y \in V^*$ such that $u = xu'y$, $v = xv'y$, and $(u', v') \in R$. The transformation to a dynamical system is as easy as in the previous example. We get a system f_G defined as follows:

$$f_G = \{(u, v) \mid \exists x, y \in V^*, u = xu'y \wedge v = xv'y \wedge (u', v') \in R\}.$$

Starting from the axiom S, the words of Σ^* reached after any finite number of derivations from the axiom form the "generated" language L_g:

$$\{S\} \xrightarrow{f_G} L_g.$$

This language is nothing but the forward attractor of $\{S\}$.

Example 9.3 (Trace language). In Chap. 4, we introduced the concept of trace language L_t as the set of all possible traces a system (X, f) can generate on a covering (α, Σ) (see Def. 4.8). This language can be seen as generated language, too. Let us show this informally. We need a predicate guaranteeing that a specific trace σ is possible or observable, possible(σ) (e.g. a fullness property, see Def. 5.24), and the trace language can be defined as $L_t = \{\sigma \in \Sigma^\omega \mid$ possible$(\sigma)\}$. The system f_T is then defined on Σ^∞:

$$f_T = \{(u, v) \mid (v = uj) \wedge \text{ possible}(v)\}.$$

Using this last system, the trace language is obtained as an attractor, too:

$$\{\varepsilon\} \xrightarrow{f_T} L_t.$$

Thus, fundamental properties of automata can be expressed as attraction properties of dynamical systems. This motivates two research directions [46, 45, 113, 317, 334]: first, a comparison of computational abilities of dynamical systems and classical computational models; second, a deeper study of connections between dynamical systems and languages, grammars and automata. The first question is the aim of this chapter.

In computability theory, *Church-Turing's thesis*, claiming that every algorithm can be realized by a Turing machine, seems to have a folklore corollary that nothing more powerful than a Turing machine exists [146]. This is not true but clearly distinctive criteria must be specified.

To illustrate our point, we concentrate on two models, apart from Turing machines: cellular automata and continuous functions. We precisely consider these models because their action-effect pairs present interesting structural particularities that constitute the main reason of their different behaviors and computational abilities.

Informally, Turing machines (TM) only produce local modifications on their tape, under the head, and the resulting effect is of course local; cellular automata (CA) produce a local change in each cell, which translates into a possible global effect; finally, continuous functions on the real numbers (CF) seem to produce a global modification at each position of the decimal expansion of numbers, which results in a possible global effect. These characteristics are summarized in Table 9.1.

Table 9.1. Pairs action-effect of computational models

Model	Action	Effect
TM	local	local
CA	local everywhere	global
CF	global	global

We embed TM, CA and CF in a general model based on the connected product, and the comparison leads to the following computation hierarchy:

$$\mathcal{TM} <_c \mathcal{CA} <_c \mathcal{CF}.$$

Some authors have already pointed out that the complexity of systems is related to undecidability issues. For example, there is no algorithm to determine the attractors of all systems [165, 224, 282]. Undecidability thus expresses the limits of systems.

Moreover, from a theoretical point of view, it is useful to propose a universal model in each class, viz. a model which is able to simulate the behavior of any other model of the same class. Universality rather expresses the power of systems.

Universality and undecidability are intrinsically related in theoretical computer science; we will see that it is interesting to extend these notions to dynamical systems in general.

9.2 Comparing Dynamical Systems

In what follows, we will compare Turing machines, cellular automata and continuous functions. To compare dynamical systems, two methods are possible: extrinsic and intrinsic. The aim of this section is to make a clear distinction between the two methods, and to present the properties used in the rest of this chapter.

9.2.1 Extrinsic Method

This method is based on simulation. Informally speaking, a system A simulates a system B if A's computations mimic B's ones. The number of steps required to realize the other one's computations can be fixed in advance or based on attraction to a "halting" state.

For example, A could need three computation steps to reach B's result after one step. It could also require a finite number of steps, depending on the input data received by B. Convergence to B's answer could be preferred, in which case A's computation could be infinite.

9.2.2 Intrinsic Method

This method is based on properties of the systems we want to compare: the languages they generate, their structure, their topological entropy, etc.

For example, let us imagine we want to prove that a system A is strictly more powerful than another system B. An intrinsic method amounts to exhibit a language or a function that A is able to compute but not B. If B satisfies a property that restricts its computations, and A is not restricted the same way, A is stronger than B.

9.2.3 Our Comparison

In what follows, we use both methods to explore and compare different models. Let us just mention them here; they are formally defined in §§9.4–9.6.

– Extrinsic method: we prove that
  – cellular automata simulate Turing machines;
  – continuous functions simulate cellular automata.
  This only entails a weak hierarchy

$$\mathcal{TM} \leq_c \mathcal{CA} \leq_c \mathcal{CF}$$

  because nothing is said about the converse simulations.
– Intrinsic method. We use the following criteria:
  – continuity w.r.t. the astronomer's metric;
  – (generalized) shift-invariance;
  – Lipschitz property;
  – shift-vanishing effect;
  – behavior w.r.t. (un)computable initial conditions.
  These discriminating properties strengthen the first hierarchy:

$$\mathcal{TM} <_c \mathcal{CA} <_c \mathcal{CF}.$$

9.3 From Locality to Globality

To compare different computational models, it is useful to formally define them in a unified way. This is the aim of this section: we recall the definitions of Turing machines, cellular automata and continuous functions, and we present a general model based on the connected product unifying them. This allows us to clarify Table 9.1 by extracting interesting characteristics that will be used later in correspondence with topological and metric properties.

9.3.1 Turing Machines

As everybody acknowledges TM as a typical universal model of computation, it is interesting to compare it to other models that are suspected to be stronger.

Definition 9.4 (Turing machine). *A Turing machine M is defined by*

$$(Q, \Gamma, \Sigma, \zeta, s, B, F)$$

where Q is the set of internal states, Γ is the tape alphabet, $\Sigma \subseteq \Gamma$ is the input alphabet, $s \in Q$ is the initial state, $F \subseteq Q$ is the set of accepting states, $B \in \Gamma - \Sigma$ is a white symbol for the tape, and $\zeta : Q \times \Gamma \mapsto Q \times \Gamma \times \{-1, 0, 1\}$ is the transition function.

A Turing machine is represented in Fig. 9.1, where the head contains a state $p \in Q$, the tape content is u, and each cell i is set to u_i.

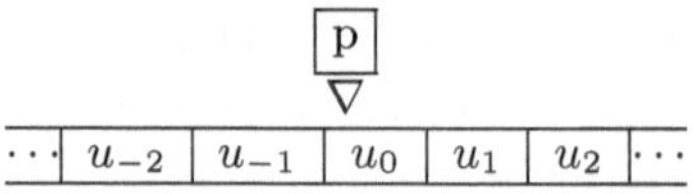

Fig. 9.1. Turing machine: u represents the tape content, and p is the internal state contained in the machine's head

At each transition, the internal state and the current tape symbol are considered: a new symbol is written on the tape, a new internal state is reached and the current position of the head on the tape is moved to the left (-1) or to the right (1), or it does not move (0).

A *configuration* c of M includes its internal state, and the whole tape content:

$$c \in Q \times \Gamma^*.$$

This state space can be embedded in $Q \times \Gamma^{\mathbb{Z}}$, in which the behavior of the TM is given by the following rules: according to the value of the internal state, and the value of the cells around zero (representing the head), modify some cells around zero, modify the internal state, and make a (null, positive or negative) shift.

A transition of the machine (head state, tape state under the head) $\xrightarrow{\zeta}$ (new head state, new tape state, shift),

$$(p, a) \xrightarrow{\zeta} (q, b, c)$$

can be rewritten as:

$$M \left(\begin{array}{c} p \\ \cdots g_{-3}g_{-2}g_{-1}.ag_1g_2g_3 \cdots \end{array} \right) = \left(\begin{array}{c} q \\ \rho^c(\cdots g_{-3}g_{-2}g_{-1}.bg_1g_2g_3 \cdots) \end{array} \right).$$

Hence, M can be considered as a dynamical system acting on the configuration space $Q \times \Gamma^{\mathbb{Z}}$. This kind of construction can be easily adapted to finite and pushdown automata.

This view of TM as dynamical system is called *Turing machine with moving tape*, i.e. TMT, in [172], because the whole tape is moved after the possible change under the head. In the same paper, the author proposes a dual way to consider TM, namely *Turing machine with moving head*, i.e. TMH. There, $Q \times \Gamma^*$ is embedded in some $X^{\mathbb{Z}}$, but one and only one cell represents the head; under the head, a slight modification can occur, and everywhere else, nothing happens. The definition given above remains valid but the viewpoint has to switch from the head to the tape: in TMH, the head moves over the tape, whereas in TMT, the tape moves under the head.

Considering TMT could lead to infinite effect domains because the last shift can entail a global modification, whereas the action domain of any TM is local, because only a finite number of cells are taken into account to compute the next global state and the effect domain is also local because only a finite number of cells can be modified at the same time, i.e. in one time step. This is the reason why we will only consider TMH in which only a few cells are modified around the head, and all other cells remain unmodified and immobile.

9.3.2 Cellular Automata

CA were defined and studied in Chap. 8 regarding attraction and composition. Let us briefly recall their definition based on the connected product (see also Def. 8.2).

Definition 9.5 (Cellular automaton). *A cellular automaton is a connected product $\otimes_R g_i$ defined on a Cartesian product $E = \times_{i \in I} X_i$ such that the local spaces are $X_i = X = \{0, 1, \cdots, k-1\}$, the neighborhood relation is $R = \{(i, \{i-1, i, i+1\}) \mid i \in I\}$, and the local transition function is $g_i = g : X^3 \mapsto X$.*

In the following, we consider elementary CA, with $I = \mathbb{Z}$ and $k = 2$. Their behavior is completely characterized by this definition. A CA is represented in Fig. 9.2. Each cell i is set to a local state u_i. The global configuration is u.

The action domain of a CA is local for every cell. The effect domain is global because every cell can be modified.

$$\cdots \leftrightarrow \boxed{u_{-2}}_{-2} \leftrightarrow \boxed{u_{-1}}_{-1} \leftrightarrow \boxed{u_0}_0 \leftrightarrow \boxed{u_1}_1 \leftrightarrow \boxed{u_2}_2 \leftrightarrow \cdots$$

Fig. 9.2. Cellular automaton: u is the global state

9.3.3 Continuous Functions

For this class of systems, the behavior in terms of local/global action/effect is less clear than in the previous cases. For instance, consider the logistic map $f(x) = 4x(1-x)$ and its effect, if any, on decimal digits of states in the unit interval:

$$0.707106781 \cdots \xrightarrow{f} 0.828427124 \cdots$$

What are the relationships between the digits, apart from the fact tey constitute two different real numbers $2^{-\frac{1}{2}}$ and $f(2^{-\frac{1}{2}})$? Is is possible to compute each digit y_i of the result $f(x)$ as a function f_i of x? If it is, how many digits of x are needed to evaluate $f_i(x)$?

As outlined by Table 9.1, the action is global and the effect is also global. This means each digit y_i of the decimal expansion of the image $y = f(x)$ of a global configuration x seems to be the image of all or many digits of x, whereas CA reduce the action-neighborhood to a bounded finite radius.

To investigate this comparison w.r.t. computability, we suspect this characteristic to be of fundamental importance.

Remark 9.6. In the rest of this chapter, we concentrate on functions that are continuous on $X^{\mathbb{Z}}$ w.r.t. the astronomer's metric, which is not equivalent to continuity w.r.t. the Euclidean distance on $\mathbb{R}$. We come back to this issue in the next paragraph.

9.3.4 General Model

Now, let us present a general connected product acting on a bi-infinite product of finite discrete spaces. This model will be specified using different instances of some parameters, and the properties of the systems so obtained will be analyzed.

We define local state spaces as finite discrete spaces, and repeat this a bi-infinite number of times via a Cartesian product. This gives:

$$\begin{aligned} I &= \mathbb{Z} \\ \forall i \in I, X_i &= X = \{0, 1, \cdots, k-1\} \\ E &= \times_{i \in I} X_i = X^{\mathbb{Z}}. \end{aligned}$$

On this global state space E, we define the general model.

Definition 9.7 (General model). *The general model is a RDS* $(E, \otimes_R g_i)$ *where*

$$R = \{(i, M_i) \mid i \in I, M_i \subseteq \mathbb{Z}\}$$

is the neighborhood relation, and

$$g_i : X^{\# M_i} \mapsto X$$

is a local transition function.

The *neighborhood* of component i is given by M_i. It contains the indices of components that can have an effect on the value of component i. We denote the cardinality of each M_i by $m_i = \# M_i$, and $m = \sup_{i \in I} m_i$. Two important parameters are the *local distance of action*:

$$\Delta_i = \max(|i - \max M_i|, |i - \min M_i|)$$

and its supremum:

$$\Delta = \sup_i \Delta_i.$$

Remark 9.8. There are particular examples of such distributed systems where the apparent action domain is infinite but its effective action domain is finite or even empty. For instance, take a constant CA, viz. where the state of each cell is left constant over time; its effective action domain cannot be considered as infinite, because no cell is taken into account to compute the next global step.

Among I, we can fix a subset $\mathbb{I}$ of effectively *active components* at a particular time-step, which means that $\mathbb{I}$ can vary during the evolution. The local action is given by g_i. All others are identities, i.e.

$$\forall j \in I \backslash \mathbb{I}, R(i) = \{i\}, g_i = \mathcal{I}_X.$$

In order to evaluate the distance between states of E, we again use an astronomer's metric (see Def. 2.53): x and y being two global states of E,

$$d_a(x, y) = k^{- \inf\{i \geq 0 | x_i \neq y_i \vee x_{-i} \neq y_{-i}\}}.$$

This metric induces the product topology on E. The more local values two states have in common, the closer they are to each other. It corresponds to our intuitive notion of distance for real numbers for example, because we are used to consider decimal representations of these numbers, in which proximity is related to the number of common decimals. Although the topologies induced by the Euclidean metric and this metric are not completely equivalent, we first examine this metric as a good candidate to generate continuous functions.

Now, we summarize the parameters of the connected product $\otimes_R g_i$ defined above:

- *action domain*: local neighborhoods M_i (m_i, Δ_i) and the greatest local distance of effect m, Δ; if all Δ_i are finite, the greatest local distance of effect Δ can be finite (e.g. $\forall i, \Delta_i = 1$) or infinite (e.g. $\forall i, \Delta_i = 2^i$); there are four cases:

$$- \forall i, \Delta_i \in \mathbb{N} \land \Delta \in \mathbb{N} \qquad\qquad (A1)$$
$$- \forall i, \Delta_i \in \mathbb{N} \land \Delta = \omega \qquad\qquad (A2)$$
$$- \exists i, \Delta_i = \omega \Rightarrow \Delta = \omega \qquad\qquad (A3)$$
$$- \forall i, \Delta_i = \omega \Rightarrow \Delta = \omega; \qquad\qquad (A4)$$

- *effect domain*: $\mathbb{I}$; we have two variants, based on the following definition:

$$\mathbb{I}(f, x) = \{i \in \mathbb{Z} \mid (f(x))_i \neq x_i\}$$

- either $\mathbb{I}$ is always finite, $\qquad\qquad\qquad\qquad\qquad (E1)$
- or it can be infinite; $\qquad\qquad\qquad\qquad\qquad\qquad (E2)$
- *local transition function*: g_i;
  - homogeneous local functions, i.e. $\forall i \in \mathbb{I}, g_i = g,$ $\qquad (F1)$
  - heterogeneity. $\qquad\qquad\qquad\qquad\qquad\qquad\qquad\qquad (F2)$

Different sets of local parameters define different instances of the connected product and, thus, different computational models. The aim of this chapter is to compare the computational abilities of these models using extrinsic and intrinsic global properties, resulting in a compositional analysis of computational properties:

$$\text{local parameters} \longleftrightarrow \text{models} \longleftrightarrow \text{global properties.}$$

Not all combinations of these properties are valid or interesting. Table 9.2 indicates by / the categories which are not considered here, and by $\emptyset$ the empty categories; we concentrate on classical models.

Table 9.2. Combinations of properties for some specific systems (TM = Turing machines, CA = cellular automata, CF = continuous real functions, GS = generalized shifts, RN = random networks)

F	E	$A1$	$A2$	$A3$	$A4$
1	1	TM, GS	/	$\emptyset$	/
1	2	CA	/	$\emptyset$	/
2	1	TM, GS	/	/	/
2	2	RN	CF	/	/

In particular, it is clear that

- $A3 - F1$ is empty because it implies $A4$;
- in some sense, $A1$ is contained in $A2$, if we consider $A2$ as the class of systems for which $\Delta \in \mathbb{N} \cup \{\omega\}$ instead of just $\Delta = \omega$;
- TM(H) and generalized shifts (GS, see [226, 225]) fall in class $A1 - E1$;
- CA are in class $A1 - E2 - F1$;
- random networks (RN, see [113]) are in $A1 - E2 - F2$, and d_a-continuous functions are in $A2 - E2 - F2$.

Are these classes sufficient or necessary characterizations of these groups of systems? Are $A3 - E1 - F2$, $A4 - E1 - F2$ and $A4 - E1 - F1$ equivalent to oracle Turing machines, assuming that an infinite action domain could be considered as an oracle? Do the last classes correspond to generalized fully-connected neural networks?

9.4 Comparison Through Simulation

Extrinsic and intrinsic methods can be used to compare computational models. In this section, we emphasize the use of extrinsic methods and we show that TM can be simulated by CA, which in turn can be simulated by CF. This leads to a weak hierarchy because CA are at least as strong as TM, and CF are at least as strong as CA. However, nothing tells us that CA are strictly stronger than TM, for instance. Yet, simulation is interesting to compare models because very often the limits of a model appear when it is embedded in a stronger one.

9.4.1 Simulation

A system A simulates another system B when every step in the computation (or evolution) of B can be performed by one or several steps of A. In general, we will consider that this number of steps is independent from the input considered. If A simulates B, then A is more powerful than B but this is only a weak result denoted by $B \leq_c A$. If one is able to show whether the converse holds or not, we have a stronger result. In the affirmative case, both systems are said to be equivalent, denoted by $A \equiv_c B$, whereas the negative case induces a strict relation between them, denoted by $B <_c A$.

Several results exist on simulation (see §9.8 for a general discussion). We consider here a general definition of the concept.

Definition 9.9 (Simulation). *Let (X, f) and (Y, g) be two dynamical systems. Then f simulates g iff there exists an encoding $\phi \in \mathcal{R}(Y, X)$ and a decoding $\psi \in \mathcal{R}(X, Y)$, such that $\exists m > 0, \forall y \in Y, g(y) = \psi(f^m(\phi(y)))$.*

Each step of g is simulated by a constant number m of steps of f. An exponent m greater than 1 allows the simulation of g by f in more than one step. The following diagram represents the homomorphic simulation:

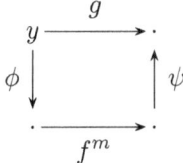

Remark 9.10. – Simulation is generally seen as a particular case of abstraction (see Chap. 4). However, this homomorphism slightly differs from the previous ones encountered so far: the decoding has to be defined from X to Y, in order to give g's image through f. A weaker simulation would be expressed as follows: $\psi'(g(y)) = f^m(\phi(y))$.

– Sometimes, another definition is given for simulation, swapping $\exists$ and $\forall$: $\forall y \in Y, \exists m > 0, \cdots$. Then a finite-time stabilization is used instead of a step-by-step simulation: the computation of f stabilizes on a fixed point or on a cycle after m iterations of f, depending on y. This is closer to intrinsic methods because it emphasizes attraction properties of each system [114, 111, 113].

9.4.2 Choice of Coding

In general, the choice of coding entails a trade-off regarding the power of encodings and decodings themselves:

– they should not be able to do too many things, because the computations have to be carried out by the model itself and not by the codings;
– they have to be powerful enough to code enough states.

In general, we consider at most a Turing computable coding, because we use simulation to compare TM with potentially stronger models. In this case, computable codings do not offer too much, as they are restricted to the suspected poorest model; they offer enough to code large state spaces; their power is well known and described.

Examples of ϕ and ψ are respectively a one-to-many relation and a many-to-one function. If ϕ is a bijective function, we can consider $\psi = \phi^{-1}$.

In the rest of this section, we use different kinds of codings to compare our different models but in the next section (§9.5), which is devoted to intrinsic comparison methods of systems, the notion of coding is not used anymore. Indeed, we consider there systems that all work on the same state space of infinite sequences of symbols, embedded into a product topology.

9.4.3 From TM to CA

The first problem to solve is the following. Given a Turing machine TM, we want to find a cellular automaton CA, an encoding ϕ and a decoding ψ such that CA simulates the Turing machine in one step: $TM(u) = \psi(CA^m(\phi(u)))$, with $m = 1$.

Our construction is not optimal but easy to understand. Equivalent results can be found in [296, 204, 168].

Encoding. We consider a Turing machine M defined as a dynamical system on $Q \times \Gamma^{\mathbb{Z}}$. We want to translate any state $(p, c) \in Q \times \Gamma^{\mathbb{Z}}$ into a state in the configuration space of a one-dimensional cellular automaton. A special symbol $*$ is added to Q, in order to keep track of the machine's head: all cells but the one representing the head will contain $*$. Any cell of the resulting automaton contains a pair (symbol, state) from $X = \Gamma \times (Q \cup \{*\})$. The configuration space of the resulting cellular automaton is a subset of $X^{\mathbb{Z}}$. More formally, the encoding is given by (see Fig. 9.3):

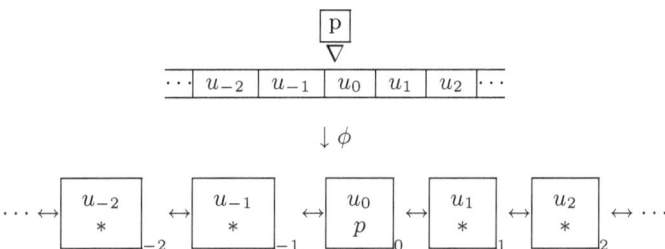

Fig. 9.3. From a TM configuration to a CA configuration

$$\phi \; : \; Q \times \Gamma^{\mathbb{Z}} \mapsto X^{\mathbb{Z}}$$
$$\text{s.t.} \quad \phi(p, c)_i = \left\{ \begin{array}{ll} (c_i, p) & \text{if} \quad i = 0 \\ (c_i, *) & \text{if} \quad i \neq 0 \end{array} \right. .$$

Using this encoding, the resulting machine becomes a TM with moving head [172], which is the only reasonable choice. Indeed, a TMT would require a global information spread over the whole lattice.

CA Local Transition Function. One step of the Turing machine, e.g. a transition $\zeta(p, u_0) = (q, a, -1)$, must give rise to an equivalent change in the CA configuration; this is depicted in Fig. 9.4.

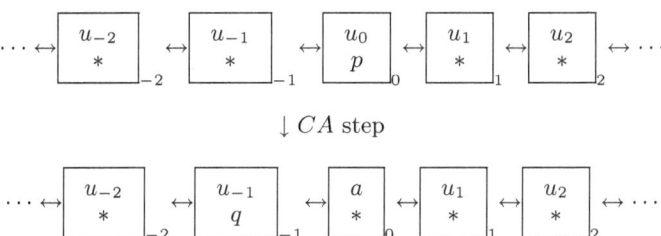

Fig. 9.4. A CA transition equivalent to a TM transition

Let us now construct the local rule of the automaton simulating the behavior of the original Turing machine. The cell containing a state different from $*$ is considered as the head; only its two neighbors have to know its current value to compute the next state of the machine. All triples $((x, *), (y, *), (z, *))$ remain unmodified. Triples with one pair (y, p), where $p \neq *$, are modified according to the transition rule $\zeta : Q \times \Gamma \mapsto Q \times \Gamma \times \{-1, 0, 1\}$ of the Turing machine. The pair (y, p) is replaced by its new value (y', q), and the shift occurs. The complete rule is represented in Table 9.3.

Table 9.3. CA local rule equivalent to the TM transition relation

triple\shift	-1	0	1
z x y * ' * ' p	z x y' * ' q ' *	z x y' * ' * ' q	z x y' * ' * ' *
x y z * ' p ' *	x y' z q ' * ' *	x y' z * ' q ' *	x y' z * ' * ' q
y z x p ' * ' *	y' z x * ' * ' *	y' z x q ' * ' *	y' z x * ' q ' *
x y z * ' * ' *		x y z * ' * ' *	

By construction, the global dynamical system $f : X^{\mathbb{Z}} \mapsto X^{\mathbb{Z}}$ resulting from the local rule presented above, is stable on a strict subset D of $X^{\mathbb{Z}}$:

$$D = \{c \in X^{\mathbb{Z}} \mid \exists \text{ a unique } i \in \mathbb{Z}, c_i = (s, p) \text{ and } p \neq *\}.$$

Indeed, the state-part of any configuration in its dynamics contains exactly one state different from $*$. All others are equal to $*$. The only cell containing a state $p \neq *$ is the head of the Turing machine.

Decoding. Based on the following functions, with $c = (c_i)_{i \in \mathbb{Z}} \in D \subseteq X^{\mathbb{Z}}$:

$$
\begin{aligned}
h &: \quad D \mapsto \mathbb{Z} \\
\text{s.t.} &\quad h(c) = k \text{ with } \forall i, c_i = (s_i, p_i) \text{ and } (i = k) \Leftrightarrow (p_i \neq *)
\end{aligned}
$$

$$
\begin{aligned}
\Pi_Q &: \quad X \mapsto Q \\
\text{s.t.} &\quad \Pi_Q(c_i) = p_i \text{ with } c_i = (s_i, p_i)
\end{aligned}
$$

$$
\begin{aligned}
\Pi_\Gamma &: \quad X \mapsto \Gamma \\
\text{s.t.} &\quad \Pi_\Gamma(c_i) = s_i \text{ with } c_i = (s_i, p_i)
\end{aligned}
$$

$$
\begin{aligned}
\Pi_{\Gamma^{\mathbb{Z}}} &: \quad D \mapsto \Gamma^{\mathbb{Z}} \\
\text{s.t.} &\quad \Pi_{\Gamma^{\mathbb{Z}}}(c) = s \text{ with } \forall i, s_i = \Pi_\Gamma(c_i)
\end{aligned}
$$

we can also define the decoding ψ (see Fig. 9.5):

$$\psi \ : \ D \mapsto Q \times \Gamma^{\mathbb{Z}}$$
$$\text{s.t.} \quad \psi(c) = (\Pi_Q(c_{h(c)}), \rho^{-h(c)}(\Pi_{\Gamma^{\mathbb{Z}}}(c))).$$

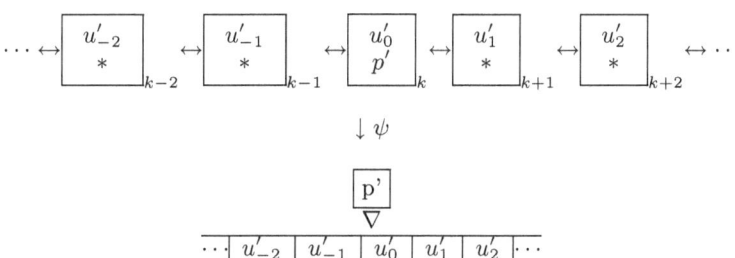

Fig. 9.5. From a CA configuration to a TM configuration

Notice that each cell of the automaton contains a pair (tape symbol, head state). In the previous formula, h gives the head position, Π_Q gives an internal state from a cell state, Π_Γ gives a tape symbol from a cell state, and $\Pi_{\Gamma^{\mathbb{Z}}}$ gives the whole tape, i.e. it removes the head symbols from a state of D. Then, ψ gives a Turing machine state (internal head state, tape state) from a global cellular state of D, shifting the head position to position 0.

Summary. Hence, any Turing machine M can be simulated by a cellular automaton f, in one CA-step per TM-step:

$$\forall y \in Q \times \Gamma^{\mathbb{Z}}, M(y) = \psi(f(\phi(y))).$$

9.4.4 From CA to CF

Now, given a cellular automaton CA, we want to find a continuous function CF, an encoding ϕ and a decoding ψ such that the function simulates the cellular automaton in one step: $CA(u) = \psi(CF^m(\phi(u)))$, with $m = 1$.

The results of this section mainly come from [227]. We use it for the sake of completeness. Another approach is described in [30]. Here we work with cellular automata spaces and real numbers from the interval $[0, 1]$, or more precisely, from a subset of $[0, 1]$, which is a Cantor set.

Every configuration of the cellular space $X^{\mathbb{Z}}$ can be transformed into a state of $X^{\mathbb{N}}$:

$$\cdots c_{-2}c_{-1}.c_0c_1c_2 \cdots \to .c_0c_{-1}c_1c_{-2}c_2 \cdots$$

where $X = \{0, 1, \cdots, k - 1\}$. This transformation ϕ_1 is a homeomorphism between $X^{\mathbb{Z}}$ and $X^{\mathbb{N}}$. Hence, we consider infinite sequences in what follows.

Every sequence of $X^{\mathbb{N}}$ can be transformed into a real number, through the following encoding function:

$$\phi_2(c) = \sum_{i=0}^{\omega} c_i \alpha^i,$$

where $c \in X^{\mathbb{N}}$, each $c_i \in X$, and $\alpha \leq \frac{1}{k}$.

In fact, if $\alpha < \frac{1}{k}$, the resulting point belongs to a Cantor set $\mathcal{D}$, whose greatest gaps are of width $1 - \alpha k$. This is important because it allows us to simulate any given cellular automaton f by a function F which is only defined on the Cantor set. Of course, this relation F can be extended to the whole interval but it has to be stable on the Cantor set, i.e. $F(\mathcal{D}) \subseteq \mathcal{D}$, because the decoding is only defined on it.

If $\phi = \phi_1; \phi_2$ and $\psi = \phi^{-1}$ are continuous functions, and keeping in mind that f is also a continuous function (see §9.5), it is possible to show that the function F defined on $\mathcal{D}$ as

$$\begin{aligned} F &: \quad \mathcal{D} \mapsto \mathcal{D} \\ \text{s.t.} &\quad F(x) = \phi(f(\psi(x))) \end{aligned}$$

is also continuous on $\mathcal{D}$. Each gap between points of $\mathcal{D}$ can be filled by simple interpolation, and we get a continuous function on $[0, 1]$.

9.4.5 Weak Hierarchy

The first simulation proves that $\mathcal{TM} \leq_c \mathcal{CA}$. Actually, neither the converse nor its negation were proved. Thus, we do not know whether $\mathcal{TM} \equiv_c \mathcal{CA}$ holds or not.

In the same way, the second simulation proves that $\mathcal{CA} \leq_c \mathcal{CF}$, and we get the following weak hierarchy:

$$\mathcal{TM} \leq_c \mathcal{CA} \leq_c \mathcal{CF}.$$

9.5 Topological and Metric Properties

The comparison based on extrinsic properties leads to a weak hierarchy of systems. This section is devoted to intrinsic comparison methods, and we face the problem as to whether the previous classes can be related to metric and topological properties? In particular, we are interested in four properties: continuity, shift-invariance, Lipschitz property, and shift-vanishing effect. The first two have been studied in [143, 273, 113] as necessary and sufficient conditions for locality of functions (class $A1 - E2 - F1$).

9.5.1 Continuity

We work in the compact metric space (E, d_a). Let us state the definition of continuity we need.

Definition 9.11 (Continuity). *A function f defined on a metric space (E, d_a) is d_a-continuous iff*

$$\forall x, \varepsilon, \exists \delta, \forall y, d_a(x, y) < \delta \Rightarrow d_a(f(x), f(y)) < \varepsilon. \tag{C}$$

Theorem 9.12. *In the general model $(E, \otimes_R g_i)$, the following equivalence holds:*

$$(C) \Leftrightarrow ((A1) \vee (A2)).$$

Proof. $\Leftarrow$We have to show that assuming $(A1) \vee (A2)$ holds, the general system f is d_a-continuous: fixing x and ε, we want to find a δ such that $\forall y, d_a(x, y) < \delta \Rightarrow d_a(f(x), f(y)) < \varepsilon$.

For any ε, we take the first n such that $k^{-n} \leq \varepsilon \leq k^{-(n-1)}$. If we prove that $d_a(f(x), f(y)) < k^{-n}$, the last inequality is a fortiori verified, which means $f(x)$ and $f(y)$ are at least equal on $\{-n : n\}$.

Now, we have to determine the action domain of this finite segment. Therefore, we define $a_m = \min_{i=-n}^{n} \min M_i$ and $a_M = \max_{i=-n}^{n} \max M_i$, and $a = \max\{|a_m|, |a_M|\}$. The action domain is $\{a_m : a_M\}$. We consider a larger domain containing this one: $\{-a : a\}$. If x and y are at least equal on $\{-a : a\}$, we have $d_a(x, y) \leq k^{-(a+1)}$. Since we want a strict inequality, we fix $\delta = k^{-a}$.

$\Rightarrow$If there exists an index i such that $\Delta_i = \omega$, then the general system f is not continuous.

Let us assume $i > 0$. For any $0 \leq k < i$, it is always possible to find a y close to x guaranteeing $f(x)$ and $f(y)$ to be equal on $\{-k : k\}$. As i has to be reached too, continuity is lost because this component depends on the whole configuration. One can thus fix arbitrarily close states x and y such that a very small distance between them influences a difference between their images at position i.

9.5.2 Shift-Invariance

Let us recall the definition of a shift: for any bi-infinite sequence x of $X^{\mathbb{Z}}$,

$$\forall i, (\rho(x))_i = x_{i+1}.$$

Definition 9.13 (Shift-invariance). *A function f defined on $X^{\mathbb{Z}}$ is shift-invariant iff f and ρ commute: $\forall x \in X^{\mathbb{Z}}$,*

$$f(\rho(x)) = \rho(f(x)).$$

Definition 9.14 (Generalized shift-invariance). *A function* f *defined on* $X^{\mathbb{Z}}$ *is* (m,n)*-shift-invariant iff* $\exists n, m \in \mathbb{Z}, \forall x \in X^{\mathbb{Z}}$

$$f(\rho^m(x)) = \rho^n(f(x)). \tag{SI}$$

Theorem 9.15. *In the general model* $(E, \otimes_R g_i)$, *the following equivalence holds:*

$$(SI) \Leftrightarrow \begin{cases} (F1) \\ \forall i, m_i = m' \\ \forall i, M_{i+n} = M_i + m. \end{cases}$$

Proof. The proof is based on the following equivalences:

$$
\begin{aligned}
& f(\rho^m(x)) = \rho^n(f(x)) \\
\equiv \quad & \forall i, (f(\rho^m(x)))_i = \rho^n(f(x))_i \\
& \qquad \because \quad (F1) \wedge (\forall i, m_i = m') \\
\equiv \quad & \forall i, g(\Pi_{M_i}(\rho^m(x))) = (f(x))_{i+n} \\
\equiv \quad & \forall i, g(\Pi_{M_i+m}(x)) = g(\Pi_{M_{i+n}}(x)) \\
& \qquad \because \quad \forall i, M_{i+n} = M_i + m \\
\equiv \quad & true.
\end{aligned}
$$

9.5.3 Lipschitz Property

The Lipschitz condition goes as follows.

Definition 9.16 (Lipschitz condition). *A function* f *defined on a metric space* (E, d) *is Lipschitzian iff* $\exists K, \forall x, y,$

$$d(f(x), f(y)) \le K \cdot d(x, y). \tag{L}$$

Remark 9.17. – Contracting functions (see Def. 2.61) are Lipschitzian with a coefficient $K < 1$.
– Condition (L) clearly implies continuity; for any $\varepsilon > 0$ it suffices to choose $\delta = \frac{\varepsilon}{K}$.

Theorem 9.18. *In the general model* $(E, \otimes_R g_i)$ *defined on the metric space* (E, d_a), *the following equivalence holds:*

$$(L) \Leftrightarrow (A1).$$

Proof. $\Leftarrow$If $(A1)$ holds, we have $\Delta \in \mathbb{N}$. Thus, if $d_a(x, y) = k^{-n-\Delta}$, we have also $d_a(f(x), f(y)) \le k^{-n}$. This last term can be equal to $K \cdot d_a(x, y)$ if $K = k^{-\Delta}$, which is constant. Thus, f is Lipschitzian.

$\Rightarrow$Equivalently, we want to show that

$$(\Delta = \omega) \Rightarrow \neg(L).$$

If $\Delta = \omega$, forall K, there exists an i such that $\Delta_i = K$. Thus, $\forall K, \exists x, y$ such that $d_a(x,y) = k^{-i-K}$ and $d_a(f(x), f(y)) \geq k^{-i}$ ($=$ if $\forall |j| < |i|, |\max M_j| < |\max M_i|, >$ if $\exists |j| < |i|$ such that $|\max M_j| \geq |\max M_i|$). Hence, we get

$$\frac{d_a(f(x), f(y))}{d_a(x,y)} \geq k^K$$

that can be arbitrarily large and f is not Lipschitzian.

9.5.4 Shift-Vanishing Effect

This last property formally expresses the pure locality of a system, that is, the finite number of local states that are modified in one computation step.

Definition 9.19 (Shift-vanishing effect). *A function f defined on a metric space (E, d_a) is shift-vanishing iff*

$$\forall x, \lim_{n \to \pm\infty} d_a(\rho^n(x), \rho^n(f(x))) = 0. \qquad (SV)$$

The following theorem characterizes the systems corresponding to this property.

Theorem 9.20. *In the general model $(E, \otimes_R g_i)$, the following equivalence holds:*

$$(SV) \Leftrightarrow (E1).$$

Proof. $\Leftarrow$For all x, we have $\mathbb{I}(f, x)$ finite. Thus, the modified states are between $\min \mathbb{I}(f, x)$ and $\max \mathbb{I}(f, x)$. By arbitrarily large shifts, the distance between x and its image vanishes.
$\Rightarrow$We prove that the contrary is false. $\neg(E1)$ means there exists x such that $\mathbb{I}(f, x)$ is infinite. Assume it is infinite to the right, that is, for positive indices: $\forall i, \exists j > i$ such that $(f(x))_j \neq x_j$. Thus, $d_a(\rho^j(x), \rho^j(f(x))) = 1$, and there is no convergence of the shifted distance to 0, i.e. $\neg(SV)$.

9.5.5 Nondeterminism

In [273, 113], hints are presented to extend Theorems 9.12, 9.15, 9.18, and 9.20 to general relations. In Def. 9.7, this amounts to change local transition functions into relations:

Definition 9.21 (General relational model). *The general relational model is a RDS $(E, \otimes_R g_i)$ where*

$$R = \{(i, M_i) \mid i \in I, M_i \subseteq \mathbb{Z}\}$$

is the neighborhood relation, and

$$g_i \in \mathcal{R}(X^{\# M_i}, X)$$

is a local transition relation.

The main problem is to transform our basic properties (continuity, shift-invariance, Lipschitz condition, shift-vanishing effect) in such a way they remain meaningful for general relations.

Before giving the extended properties, we need the notion of cluster point and a lemma guaranteeing their existence for any sequence of global states [273].

Definition 9.22 (Cluster point). *A global state x is the cluster of an infinite sequence $(x^i)_i$ of global states x^i iff every finite set of local states which agrees with x also agrees with x^i for infinitely many i.*

Lemma 9.23. *Every infinite sequence of global states has a cluster point.*

Now, we briefly present these extended properties, and leave the corresponding extended proofs to the reader. The Hausdorff metric h used here is based on d_a.

- *r-Continuity:* if the sequence of global states $(x^i)_i$ has only one cluster point x^ω, then $x \in f(x^\omega)$ iff there exists a sequence $(y^i)_i$ whose cluster point is x (always exists thanks to Lemma 9.23) such that $\forall i, y^i \in f(x^i)$;
- *r-Generalized shift-invariance:*

$$y \in f(x) \text{ iff } \rho^n(y) \in f(\rho^m(x));$$

- *r-Lipschitz property:* $\exists K > 0$ such that $\forall x \neq y$,

$$h(f(x), f(y)) \leq K \cdot d_a(x, y);$$

- *r-Shift-vanishing effect:*

$$\forall x, \forall v \in f(x), \ \lim_{n \to \pm\infty} d_a(\rho^n(x), \rho^n(v)) = 0.$$

The next condition has to be added to each theorem, in order to extend them to relations; it is called *independence condition*: if there exists a global state x, a sequence of indices $(i_k)_k$, and a sequence of local states $(r_k)_k$ such that $\forall k, \exists y^k$ such that $y^k \in f(x)$ and $\Pi_{\{i_k\}}(y^k) = r_k$, then there exists a global state y such that $y \in f(x)$ and $\forall k, \Pi_{\{i_k\}}(y) = r_k$.

9.5.6 Summary

Let us summarize the properties studied in this section (see Table 9.4), and relate them to the models introduced in §9.3 (see Table 9.5).

Using the informal statements of Table 9.1 and the definitions of Turing machines (Def. 9.4) and cellular automata (Def. 9.5), we conclude that

- continuous functions naturally correspond to condition (C);
- cellular automata, by definition, satisfy conditions (C) as particular case of our general model, (SI) and (L);
- Turing machines, by definition and keeping in mind that we concentrate on the TMH version of the model, satisfy all properties of CA and (SV). In particular, observe that (F1) and, thus, (SI) would not be verified as such in the moving-tape version of TM.

Table 9.4. Summary of topological and metric properties

	Property	Characterization	Theorem
(C)	continuity	$(A1) \vee (A2)$	9.12
(SI)	shift-invariance	$(F1) \wedge (\forall i, (m_i = m')$ $\wedge(M_{i+n} = M_i + m))$	9.15
(L)	Lipschitz property	$(A1)$	9.18
(SV)	shift-vanishing	$(E1)$	9.20

Table 9.5. Topological and metric characterization of computational models

model\property	(C)	(SI)	(L)	(SV)
TM	×	×	×	×
CA	×	×	×	
CF	×			

Remark 9.24. Cellular automata are shift-invariant on $X^{\mathbb{Z}}$ whereas Turing machines verify this property on a strict subset D of $X^{\mathbb{Z}}$ (see §9.4), the set of "legal" configurations, with exactly one head.

The intrinsic comparison based on topological and metric properties extends results of [143, 273] to our model: basically, the authors proved the equivalence between CA and d_a-continuous ρ-invariant functions on $X^{\mathbb{Z}}$. We added new properties (generalized shift-invariance, Lipschitz condition, shift-vanishing effect), in order to characterize the three models in one general framework. In [323], a similar definition of continuous functions is given.

9.6 Computability of Initial Conditions

In this section, we go on comparing our systems using intrinsic properties. In the previous section, we focused on topological and metric properties. Here, we turn our attention to the behavior of the different models when they start from computable or uncomputable initial conditions.

The set of possible configurations is $X^{\mathbb{Z}}$, that we decompose into two subsets: the (Turing) computable states $\mathbb{C}$ and the uncomputable ones $\mathbb{U}$. We add "(Turing)" because we want to build a hierarchy of systems where some systems are strictly more powerful than TM. A precise meaning of computation is thus required, we take the classical view of computability theory extended to uncountable domains.

Definition 9.25 (Computability). *A configuration, i.e. a global state x of $X^{\mathbb{Z}}$, is computable iff there exists a TM such that for each index n, the corresponding local value x_n can be obtained in finite time from this machine.*

We want to show that some classes of systems are stable for computable or uncomputable initial conditions.

Proposition 9.26. *Any TM M is stable on $\mathbb{C}$ and $\mathbb{U}$, i.e.*

$$M(\mathbb{C}) \subseteq \mathbb{C} \text{ and } M(\mathbb{U}) \subseteq \mathbb{U}.$$

Proof. It is clear that a local change in a sequence cannot modify its status. Under the action of a TM, the sequence thus remains in the same class.

Proposition 9.27. *Any CA M is stable on $\mathbb{C}$, i.e.*

$$M(\mathbb{C}) \subseteq \mathbb{C}.$$

Stability on $\mathbb{U}$ is not always true.

Proof. Since every local change a CA can produce in a state is computable and only depends on finitely many local states, any global computation starting from a computable state remains in the same class.

Conversely for uncomputable global states, the result of a global computation can belong to computable or uncomputable states. Take two examples: the constant global function applies everything to the same computable global state, i.e. $M(\mathbb{C} \cup \mathbb{U}) \subseteq \mathbb{C}$; the inverter, mapping a local 0 to a 1, and every local 1 to 0, is stable on $\mathbb{U}$, i.e. $M(\mathbb{U}) \subseteq \mathbb{U}$.

Proposition 9.28. *Any CF M is stable on $\mathbb{C}$, i.e.*

$$M(\mathbb{C}) \subseteq \mathbb{C}.$$

Stability on $\mathbb{U}$ is not always true.

Proof. The same arguments as in Prop. 9.27 can be applied to this class of systems.

9.7 Hierarchy of Systems

In §9.4, we compared our systems using extrinsic methods. In §§9.5, 9.6, we used intrinsic comparison methods. In this section, we summarize our analysis. The comparison appears in Table 9.6.

Table 9.6. Comparison of computational models

Properties	TM	CA	CF
Local function	homog.	homog.	heter.
Effect domain	finite	infinite	infinite
Greatest loc. dist. effect	$\in \mathbb{N}$	$\in \mathbb{N}$	$\in \mathbb{N} \cup \{\omega\}$
Simulation		TM	CA
Continuity	yes	yes	yes
Shift-invariance	yes	yes	no
Lipschitz	yes	yes	no
Shift-vanishing	yes	no	no
Stable on $\mathbb{C}$	yes	yes	yes
Stable on $\mathbb{U}$	yes	?	?
Comp. time on $\mathbb{U}$	infinite	finite	finite

Lipschitzian systems are evidently continuous, and shift-invariant systems are particular cases of generalized shift-invariant systems.

Thus TM are strictly contained in CA, which in turn are strictly contained in continuous real functions, which entails the strict hierarchy:

$$\mathcal{TM} <_c \mathcal{CA} <_c \mathcal{CF}.$$

Of course, the three classes $\mathcal{TM}$, $\mathcal{CA}$, and $\mathcal{CF}$ have the same computational power if at least one of the following assumptions is considered:

- computations can take infinite time;
- initial conditions have to be finite sequences, i.e. they belong to finite states of $\mathbb{C}$;
- computations can be approximated.

Otherwise, their exist clear distinctions between these models, as between them and noncontinuous ones.

The main drawback of all this is that CA and continuous real functions (CF), just as Turing machines, have many undecidable properties related to their long-term behavior: attraction basins, limit languages, etc. Actually, these properties are often equivalent to the halting problem of a corresponding Turing machine, which is undecidable. This undecidable problems can be seen as a much stronger complexity than classical chaos, based on sensitive dependence on initial conditions (see §9.8.3). Nevertheless, the challenge faced so far is to isolate classes of dynamical systems, and composition rules, allowing us to prove some results for specific systems (see Chap. 6).

The main positive result is that, since any Turing machine can be simulated by a cellular automaton, which in turn can be simulated by a continuous function, and since Turing machines are computationally universal, both CA and CF are universal, too. Of course, they are universal in the sense of sequential processes, but CA and CF are stronger, as we have shown in the previous sections.

9.8 Discussion

This last section summarizes the main ideas of the chapter, discusses related work, and states some open problems.

9.8.1 Composition and Computation

In this chapter, we focused on a comparison of three dynamical systems, namely Turing machines, cellular automata, and continuous functions, regarding their computational abilities.

Again, a compositional analysis was used successfully: we presented a general model based on the connected product, that we instanciated to TM, CA and CF, thereby permitting a systematic comparison of their properties.

In some cases of intrinsic properties, the compositional analysis can be expressed elegantly. Global properties G are continuity, Lipschitz conditions, etc.; individual properties I are Ai, Ej, etc. For example, Theorems 9.12, 9.15, 9.18 and 9.20 can be rephrased as equivalences between global and individual properties:

$$G(\otimes_R g_i) \Leftrightarrow \wedge_i I(g_i).$$

Although the resulting hierarchy of systems presented above is original, a number of conclusions are not, but confirm known results.

The computational model implicitly assumes that infinite sequences can be computed in finite time. This can be discussed and other assumptions would lead to different conclusions. In §9.8.3, we refer to a theory where only finite approximations of infinite sequence are computable in finite time.

9.8.2 Further Work

It is possible to build a universal Turing machine, i.e. able to simulate any other Turing machine. Of course, there exist a cellular automaton and a continuous real function corresponding to this universal Turing machine. But, is it possible to build a universal cellular automata, able to simulate any other CA, and a universal CF, able to simulate any other CF, in one or more dimensions? The first question is answered positively in [74], for CA of the same dimension (14 different local states are needed). The second answer is conjectured to be negative, unless only partial recursive functions are considered [303].

Furthermore, it seems interesting to find new intrinsic properties permitting a deeper characterization of computational models. New models could also be investigated, like DNA-computers [256], quantum computers [87, 304], or quantum cellular automata [126, 214].

9.8.3 Related Work

Comparison of Dynamical Systems. In this chapter, we mainly presented a unified view of classes of systems, that are generally separated. However, some authors have contributed to the approach. More precisely, extrinsic and intrinsic methods have already been used to compare systems. Simulation has been extensively used, with many different definitions leading to different results [68, 182, 183, 164, 166, 167, 169, 168, 226, 227, 229, 282]. Intrinsic properties usually concern language-theoretic issues, related to languages that can be generated or recognized by systems [113, 228, 283]. Ergodic theory and topology have also been used to characterize some systems [165, 168, 172, 143]. Finally, our §9.5 is mainly inspired by [273], where the author proposed a similar characterization of cellular automata.

Computability in Spaces of Infinite Objects. Using our general model, we assume that systems are able to compute the image of an infinite sequence in one shot, i.e. in finite time. This is conceivable for Turing machines but less obvious for other systems. In particular, in mathematics, one does not think about global configurations and our general model when using continuous functions on real numbers. Numbers are considered globally, without any reference to a particular machine computing the function. Only a few authors use infinite machines like CA and neural networks as computational model [113]. On the other hand, type-2 computability theory, developed in [323], extends classical (finite) computability theory to domains of infinite objects by successive Turing-computable approximations. Actually, one of his results states the equivalence between continuous functions and type-2 computable ones. The result of our Prop. 9.28 is thus not surprising.

Simulation. We cited several results on simulation here above. It is interesting to point that simulation is also very important in the theory of programs: refinement, abstraction and simulation are different words for the same concept of homomorphism between systems. For instance, see [19, 81, 76, 222] for applications in specification and verification of concurrent processes, and [272, 220] for general considerations on simulation of dynamical systems. More generally, as simulation can be considered as a particular case of abstraction, we refer the interested reader to references cited in Chap. 4.

Relating Computability to Complexity. In [302], the author introduces four classes of chaos depending on evolution and initial values generating the complexity: (I) computable evolution of CA with uncomputable initial values, (II) uncomputable evolution with computable initial values, (III) uncomputable evolution with uncomputable initial values, and (IV) computable evolution with computable initial values. To get a CA strictly more powerful than any TM, one requires uncomputable initial values. Thus, chaos I lies in a space where TM and CA are no longer equivalent, whereas chaos IV can appear in the behavior of CA as well as in TM.

Finally, let us mention interesting results on dynamical complexity as undecidability results. In [224, 226, 225], the author explains that chaos is not complex as compared to undecidability. Chaos is often reduced to sensitivity to initial conditions, where any arbitrarily small perturbation of initial conditions eventually entails a dramatic modification in the long-term behavior. Undecidability is worse because even with infinite precision, a precise result is unreachable. For example, a general Turing machine is so "sensitive to initial conditions" that it is simply impossible to know in advance whether the machine will halt or not when starting from a given initial condition on its tape, to which part of its configuration space this configuration will be attracted.

10. Epilogue: Conclusions and Directions for Future Work

Entre l'ordre et le désordre
règne un moment délicieux...

Paul Valéry

L'aventure humaine s'étale
entre la lisière du temps et de l'espace
comme une longue écharpe
d'innombrables mailles codées
depuis la nuit des temps
aux futurs déjà préparés.
L'homme aussi décrit son histoire
dès l'aube de son éveil
sur les rubans de son cerveau
les souvenirs se tissent
en une invisible écharpe
protégeante et pesante.
Chaque jour nous écrivons ensemble
une page dans le livre de l'Aventure
et la page se fera maille
préparant la maille suivante
dans la grande écharpe...

Effe Friede Voet

In this monograph, we developed a compositional analysis of dynamical and computational properties of iterated relations, i.e. discrete-time relational dynamical systems, and we used it, together with abstraction, to analyze a number of typical systems.

The approach is based on the ideas that composing individual analyses of given systems can be easier than studying a global system composed of these subsystems, and that the feasibility of these analyses depends on well-chosen abstractions.

F. Geurts: Abstract Compositional Analysis of Iterated Relations, LNCS 1426, pp. 243-255, 1998.

We defined dynamical systems as closed relations on compact metric spaces, whose dynamics was generated using a general transfinite iteration scheme. Structured systems were obtained by means of composition operators, and the abstract observation of dynamics was set up in order to cope with more realistic descriptions of evolutions.

Dynamical properties concern the complementary notions of invariance and attraction. The structure of invariants strongly influences the dynamics of systems: the richer they are, the more complex systems behave. Using fullness and atomicity, we characterized behavioral complexity, and Floyd- or Lyapunov-like sufficient criteria were proposed for attraction.

Classical dynamical systems were analyzed by composition and abstraction: Smale-horseshoe map, Cantor relation, logistic map. A family of formal systems also benefited from the compositional and abstract analysis: paper-foldings. This is the case of cellular automata, too: we classified their behaviors, the resulting classes were structured, and the most complex one was fruitfully analyzed by composition.

A general structural source of complexity was identified, using these techniques: complex behaviors arise from the composition of several dynamically compatible systems attracting the space to different invariants.

Computational properties rather concern evolutions of systems as input-output relations. Three systems were embedded in a general model: Turing machines, cellular automata, and continuous functions. Their extrinsic (simulation) and intrinsic (topological, metric and computability arguments) computation-based properties were analyzed by composition, which led to a computation hierarchy.

This last chapter starts with a summary of the main contributions and related references of this monograph (§10.1); then, we develop directions for future research (§10.2); finally, we propose a last visit of the Garden of Structural Similarities (§10.3).

10.1 Contributions and Related Work

Motivated by successful compositional results in program theory, we analyzed dynamical and computational properties of dynamical systems in a compositional way, in order to understand better how complexity arises in such systems. We hope this monograph at least partially contributes to this objective. We unified notions from dynamical systems, cellular automata, program and computability theory. Moreover, by compositional analysis, we studied important classes of complex dynamical systems.

In this section, we systematically review our main contributions, by comparison with existing ideas we based our developments on.

The first subsection mainly recalls the mathematical framework based on existing results that we adapted to our needs. The second one contains our principal contributions in compositional analysis, as well as their applications.

The references cited below are not exhaustive; many other ones can be found in the previous chapters. However, they indicate the ideas, techniques and results closest to our own work.

10.1.1 Mathematical Framework

Iteration of Relations. Relational dynamical systems are introduced in [9]; however, after the introductory chapters, the author restricts himself to functions. Set-transformers are commonly used in fractal theory [328, 159, 140, 325]. The equivalent notion of predicate-transformers is classical in program semantics [91, 93, 246, 39].

The advantage of working with relations instead of functions is twofold: it encompasses nondeterminism, and entails a homogeneous mathematical framework.

In conjunction with iterations, a notion of convergence was needed: we used lattice-fixpoint theorems [306, 197]. Convergence in at most ω steps is equivalent to continuity of set-transformers; relaxing this condition to monotonicity can require transfinite iterations [70, 71]. Bounded-nondeterminism was presented as a necessary and sufficient condition for continuity in [284]. We proved that any closed relation defined in a compact metric space verifies a slightly stronger version of continuity without bounded-nondeterminism.

The evolution of systems is obtained by successive iterations of relations or as the set of all possible trajectories. This second equivalent version is close to the well-known notion of trace in the semantics of parallelism [216, 189, 309, 89].

Iteration of Composed Relations. Compositionality is widespread in computing science and logic. Sequential programs [91, 150], parallel programs [245, 1, 67], and processes [151, 222, 79], can be studied by composition on the basis of the definitions of basic systems and composition means.

Our composition operators mainly come from [245]. Our connected product is equivalent to the database operation "join" of [53], where it is used in the context of cellular automata.

The elementary compositional results presented in Chap. 3 concern one-step evolutions (equivalent statements can be found in [245]) and infinite trajectories (treated in process algebras, e.g. [79]).

Abstract Observation of Dynamics. When trajectories are observed and abstracted, the resulting notion of trace is very similar to traces in parallelism semantics, and to coarse-grained observations in symbolic dynamics [139, 202]. It is also close to the labeling of transition systems [151, 222, 19]. Here, we integrated these various views in a single common framework.

Dynamical Properties: Invariance and Attraction. Invariance and attraction are classical notions of temporal logic [99, 98, 210] and both dynamical systems and program theory [9, 326, 91]. In ergodic theory, two notions characterize the structure of invariants [248, 196]: exactness means that all abstract traces are possible, (weak) generation limits the number of invariant states corresponding to given traces. These notions correspond to fullness and atomicity [286].

Basically, Chap. 5 refines and generalizes definitions, results and criteria presented in [286, 116, 290].

10.1.2 Compositional Analysis

Basic compositional results are presented in Chap. 3, about one-step evolutions and sets of infinite trajectories of composed systems. Here, we focus on the compositional analysis of dynamical and computational properties relying on the dynamics of composed systems.

As we said earlier, compositionality is a classical idea. However, in dynamical systems theory, only a few compositional results exist, which motivated our study inspired from related techniques in program theory and process algebras. Moreover, although it is generally not hard to express systems and properties in a compositional framework, *proving results* by compositional analysis demands a much more technical work.

In Chap(s). 6 and 7, we presented a systematic and general analysis of systems based on free products, unions, and sequential composition; less results have been obtained w.r.t. intersection, negation, and connected product.

On the other hand, in Chap(s). 8 and 9, a particular attention was devoted to connected product, whose inherent difficulty deserved a specific treatment in order to discover interesting results.

Compositional Dynamics. The dynamical properties of systems (invariance, attraction, structure of invariants) were analyzed by composition in a systematic way, for each composition operator introduced before. This led to the results of Chap. 6. Union deserved a special attention, in particular regarding the structure of its invariants: we adapted results of [328, 159, 140] to our relational framework, in terms of fullness and atomicity of invariants. Using this analysis, we gave a general way to generate complexity from elementary systems.

Case Studies in Compositional Dynamics. We systematically rederived known results about the complex dynamics of classical dynamical systems based on sequential composition (iteration), product, and union: Smale-horseshoe map, Cantor relation and logistic map. Our compositional analysis of these systems is shorter and clearer than classical approaches offer [88, 326].

In the case of paperfoldings, [80, 219, 83, 13, 14], the compositional analysis led to the following result: this system has a Cantor-set invariant on which the dynamics is chaotic.

Cellular Automata: Classification and Composition. Cellular automata are extensively studied in the literature. These systems are nice examples of connected products.

We proposed a classification refined from [49, 102], that we formalized and structured in order to characterize the most complex behaviors, and to keep its definition as close as possible to the intuition. Indeed, previous classifications were generally informal [330] or counterintuitive (e.g. [41] classifies shifting behaviors as complex).

Then, we studied a specific case of disjunctive composition. By compositional analysis, and using additional complexity measures, we compared this system to local and global unions. This led to the reinforcement of a conjecture proposed in [49, 101], on the complex behavior of cellular automata obtained by disjunction of symmetric shifting behaviors.

Compositional Computability. Three models were studied: Turing machines (TM), cellular automata (CA), and continuous functions (CF), as instances of a general system based on the connected product. We established a strict hierarchy between these systems, from TM to CF, using extrinsic and intrinsic means of comparison.

Extrinsic comparison methods based on simulation are natural in the field of computability, as well as in dynamical systems theory. In particular, our TM-to-CA simulation confirms equivalent results [296, 204, 168]; the second simulation (CA-to-CF) comes from [227].

Intrinsic comparison methods based on topological and metric properties are studied in [143, 273, 323], where the authors define continuous functions, cellular automata as continuous shift-invariant systems, and propose extensions to encompass nondeterministic systems. We added the Lipschitz condition and the shift-vanishing effect, and we generalized shift-invariance, in order to characterize Turing machines, cellular automata and continuous functions as instances of the same general model. Finally, we compared these systems w.r.t. computability of initial conditions.

Again, the compositional analysis we proposed appears quite useful, since intrinsic criteria are proved equivalent to individual properties of components of the general connected product.

10.2 Directions for Future Research

This monograph focused on theoretical aspects in the field of dynamical systems: we developed a compositional analysis of their dynamical and computational properties, using abstraction techniques. Moreover, our approach revealed useful to understand better how complexity arises in a number of typical dynamical systems.

Admittedly, our case-studies remain academic, and composition-in-the-large is far from being solved. In this section, we propose research directions which should be seen as challenges and benchmarks for the future of compositional and abstract analysis of systems.

By this work, we hope both positive and negative results could emerge. Indeed, it is mandatory to validate or invalidate ideas and techniques, and to evaluate the effective usefulness of abstraction and composition regarding real applications.

10.2.1 A Patchwork of Open Technical Issues

Let us first sum up important technical issues that have already been mentioned in the previous chapters:

- considering other times structures (continuous, hybrid, and more abstract ones);
- deeper studying the connected product and adding other composition operators, including functional fixpoint equations;
- introducing new abstractions of dynamical and computational properties such as Perron-Frobenius operators, invariant measures, and local structure theory;
- refining the attraction-based classification of cellular automata;
- strengthening the compositional analysis of computational properties of systems.

10.2.2 Fractal Image Compression

The first steps in compositional analysis of dynamical systems were carried out in fractal theory, where union was introduced to compose elementary systems so as to obtain set-based functions called "iterated function systems" [328, 159, 140, 28, 325].

Now, after a long contemplation period of beautiful fractal pictures [208, 258], fractal theory has become a source of effective tools for image compression [29, 100]. Instead of transmitting a chicken, the original egg is sent. In other words, given a bitmap image, the closest possible fractal is considered, and the compact, formal description of the dynamical system generating this fractal as attractor is sent. Provided the original image is sufficiently self-similar, the process entails a dramatic reduction in time or space and, thus, in money.

In some cases, more resources are needed to compress a global image than the sum of its components taken individually, and a better compression rate can be obtained after decomposition of the original picture. The famous "collage theorem" [28], based on the contraction-mapping theorem applied to unions of systems, is the main compositional result in this direction.

Recent publications have shown how mathematical morphology can help in structuring and analyzing images: mathematical composition operators are defined in order to build structured images from various objects and forms [280, 144, 275].

Composition as developed in this monograph seems thus appropriate to further develop this approach. Operators from mathematical morphology could be expressed in terms of our composition operators, and irreducible ones should be investigated w.r.t. dynamical properties. Hopefully, this could improve or generalize a productive and now widespread technique in image processing.

10.2.3 Distributed Dynamical Optimization

Major challenges in distributed systems are the systematic and formal verification and construction of algorithms, possibly taking probabilistic evolutions into account. In particular, formalizing relationships between structural information (topology, labeling) and complexity (time, space) represents a promising research stream in the field [108].

Our perspective is the development of robust, efficient, distributed optimization algorithms that could be able to reorient their search "on the fly", as initial conditions are modified (evaluation function, constraints, data).

When classical optimization algorithms are embedded in dynamical environments, their behavior can change dramatically, and most of the time, no solution whatsoever results. Packet routing on networks, real-time planning or scheduling, multi-routing, garbage collection are typical applications where dynamical environments and perturbations must be dealt with. Mobile telecommunications provide an important example: in general, the system is often modified before any routing algorithm can compute a globally optimal solution.

We propose to model such a problem so that its initial conditions correspond to parameters of a distributed dynamical system whose attractors are the solutions, and the global optimization emerges from individual properties of distributed agents of the system.

Several aspects must be investigated to reach the objective: composition as structuring means to build distributed systems, and emerging self-stabilization to allow reorientation despite perturbations; probabilistic transitions as suitable modelization of environment events; higher-order dynamics to describe how parameters evolve; and design of attraction-based systems. We discuss these various aspects in the following sections.

Let us mention some works that already lead to robust adaptive distributed solutions. Their inspiration sources are collective techniques exhibited by social insects, computational ecologies, and collective intelligence [65, 95, 158, 188, 198, 270]. Moreover, classical on-line algorithms, artificial intelligence techniques (neural networks, genetic algorithms, etc.), as well as nonadaptive general resolution schemes (divide and conquer, global search, local search, etc.) should be extended to dynamical systems in order to improve the resulting framework [243, 298, 295, 294].

10.2.4 Distributed Systems and Self-Stabilization

A distributed system is composed of communicating agents that cooperate in order to solve a common problem. The global state of a distributed system is obtained by simple juxtaposition of all individual states. Although individual behaviors can be characterized by local properties, the global behavior does not result from their simple combination. Indeed, the behavior of the whole system is obtained by dynamical composition of local effects, which entails global properties that remain invisible locally and statically. This phenomenon is called *emergence*.

All distributed systems undergo emerging properties. However, as the evolution leads the system to a global objective, and an organized structure emerges from a random initial condition, there is *self-organization*. Moreover, when the order is preserved despite perturbations of the system, one speaks of *self-stabilization*.

Self-stabilization is defined by two conjugate properties: convergence (attraction, liveness, termination) and closure (invariance, safety), which sustain the analysis of many families of programs and dynamical systems [90, 278, 123, 192, 15, 290, 287].

Important examples where the principal emerging property is self-stabilization are: distributed algorithms for leader election, clock synchronization or process wake-up [6, 238, 269], cellular automata [331], and neural networks [121, 148].

The compositional analysis of systems we developed in this monograph contains four basic aspects useful in the study of self-stabilizing distributed systems: composition, abstraction, invariance, attraction. Such a compositional analysis of distributed systems has already proved useful [1, 61, 62, 67, 288].

The tools and results we have presented in this monograph concern dynamical systems in general, and could be adapted and further developed to analyze distributed systems in particular, or systems combining classical programs and dynamical systems.

10.2.5 Probabilistic Systems and Measures

Probabilities are introduced in programs to accelerate them or even sometimes to give a solution while no deterministic one could exist [131]. In dynamical systems (e.g. power plants), probabilities can help in modeling the environment, when no deterministic law is known but some statistical facts are available. In general, deterministic or nondeterministic transitions between states or sets of states are replaced by probabilistic transitions.

Again, the predicate-transformers calculus have been extended so as to include probabilistic transitions [234]. Some authors have also integrated parallelism and probabilities in a common framework [147, 271, 17].

Set-transformers have been extended to probabilities in fractal theory: measures are considered instead of sets [159, 160, 28].

In the theory of dynamical systems, pointwise evolutions are not powerful enough to follow the complexity of some systems. Measures offer a better description, or abstraction, means. In particular, Perron-Frobenius operators, as well as their high-dimensional equivalent, i.e. the local structure theory [137], define the evolution of probability measures, and invariance and attraction are again used at that level [196],

A very important extension of our work should include these operators on measures: we should analyze the effect of (composed) relational dynamical systems on measure spaces.

Then, a second lifting, or abstraction again, could consist in working at the level of probability distributions instead of measures generated by these probabilities. What are the effects on invariance and attraction, on complexity, on computations, when composition of dynamics is considered at the level of probability distributions?

10.2.6 Higher-Order Systems, Control, and Learning

Any dynamical system is based on three components: a space where things happen, a relation showing how things happen, and a time structure determining when things happen. To this triple, a fourth component is sometimes added: parameters, which can be seen as an "active" part of the space where things happen. Actually, we distinguish them from "passive" variables because their goal is different in that they rather control the dynamics instead of just resulting from the evolution of systems. Moreover, their own evolution is often determined by an upper controlling system, running over the first, controlled one.

Changing the value of parameters in order to better fit a specified objective is called *learning* (from the point of view of the controlled system) or *teaching* (from the point of view of the controlling system). This entails a behavioral modification of the controlled system, called *bifurcation*.

But who controls the controllers? Who teaches the teachers? Control parametrization can be extended to several embedded levels. At each level, there is a controlled system, and its tutor or upper system changing its parameters is a *higher-order controlling system*. The evolution of the higher-order system happens in a space of systems, and is called *higher-order dynamics*.

Thus, instead of considering systems acting on "classical" discrete or continuous state spaces, we could also define them on richer state spaces containing dynamical systems. This reminds of the use of functions as first-class citizens in functional programming, as well as objects, composed of states and functions, in object-oriented programming.

We postulate this view could help in developing new learning techniques, or at least in better understanding the dynamics of existing learning algorithms, in terms of invariance, attraction and computations. This raises the following questions:

– How to get effective convergence, in finite and small amount of time?
– How to get structural stability, or robustness?

Neural networks [156, 157, 223, 121, 173, 142, 38] and genetic algorithms [153, 298, 184, 185] are two artificial intelligence techniques essentially used in optimization, classification, pattern recognition and approximation. They often prove useful when not much information is available on data to be treated. Both techniques can be seen as parametrized dynamical systems, and learning consists in fine-tuning their parameters in order to solve a particular problem. Learning algorithms are thus crucial in these fields.

Of course, this can be generalized to more than two levels. Nowadays, it is not rare to have several systems embedded to solve a particular problem. For instance, a genetic algorithm searches for the best family of neural nets able to solve a given classification problem. In this case, the set of objects is a set of neural nets, that is a set of parametrized dynamical systems.

Many cases of learning, including the supervised-unsupervised distinction, can be modeled this way. Each teacher is a higher-order system that works on systems and parameters. It receives as inputs the learning system, and the information to teach, and it produces as output a set of parameters that determine the behavior of the lower-order system. Attraction, invariance, and other properties can be studied at higher-order levels using the tools introduced in previous chapters.

10.2.7 Design of Attraction-Based Systems

Program development consists in building a correct program from a formal specification of what it is intended to do. Theoretically, a specification corresponds to the description of a language an automaton should recognize. The development of a program then corresponds to building the automaton. Using the terminolgy used in dynamical systems theory, the language corresponds to the attractor of the automaton.

This classical view can be extended to dynamical systems: given the specification of a problem, design a system whose dynamics solves the problem. Stated otherwise, given a set of states in a state space, elaborate a system whose attractor precisely is this set of states:

$$\text{specification} \xrightarrow{\quad \text{design} \quad} \text{program}$$

$$\text{viz.} \quad \text{attracting set} \xrightarrow{\quad \text{design} \quad} \text{dynamical system.}$$

In control and hybrid systems theory, recent advances combining techniques from both dynamical systems and program theory have already proved useful [187, 16, 288]. Similarly, cellular automata have been used as building blocks of structured systems, in order to derive original solutions of well-known problems in distributed algorithms (synchronization of processes, iterative generation of fractals, etc.) [215, 113].

The advantages of extending software design to dynamical systems are numerous [44]. Fault-tolerance, stability, and stabilization properties are fundamental in real-time systems for instance, and they are naturally expressed as attraction and invariance conditions in the context of dynamical systems [278, 123]. This does not mean that the design of such properties in the context of dynamical systems is immediate. Actually, they often remain quite hard to derive using current mathematical methods of program development. Hence, the techniques and criteria developed for dynamical systems should be exploited in this direction.

The challenge is to derive efficient dynamical systems from specified attraction and invariance properties. This would amount to define a design method, a language in which properties are expressed, and a logic in which a constructive reasoning can be conducted.

Compositionality is to dynamical systems what modularity is to software design. Hence, we think and hope that the compositional analysis of systems developed in this monograph, together with the abstraction principle also applied intensively, could be used as a framework for the systematic design of attraction-based systems.

10.3 The Garden of Structural Similarities

> Symmetry,
> as wide or as narrow as you may define its meaning,
> is one idea by which man, through the ages,
> has tried to comprehend and create
> order, beauty, and perfection.
>
> Hermann Weyl

Our journey in the abstract compositional analysis of iterated systems is best summarized as the discovery of a Garden of Structural Similarities[1], as a unified quest of structural affinities, similarities and symmetries, that is, homomorphisms. Indeed, through the sequence of chapters, we met many different types of these mathematical structures that not only constitute the cement of the framework and toolbox we developed, but also highlight the path to a better understanding of dynamical and computational properties of systems. Let us recall the three families of homomorphisms we used in the monograph:

---

[1] This beautiful expression is due to Michel Sintzoff.

– *composition homomorphisms* for dynamics, invariants, attractors, and computations of systems;

$$
\begin{array}{ccc}
S_i & \xrightarrow{\ I\ } & I(S_i) \\
\star \downarrow & & \downarrow \diamond \\
\star_i S_i & \xrightarrow[\ G\ ]{} & \cdot
\end{array}
$$

– *abstraction homomorphisms* for trace-parametrized invariant structures, symbolic dynamics, and extrinsic simulation properties;

$$
\begin{array}{ccc}
X & \xrightarrow{\ f\ } & X \\
\mathcal{Z} \downarrow & & \downarrow \mathcal{Z} \\
Y & \xrightarrow[\ g\ ]{} & Y
\end{array}
$$

– *compositional complexity homomorphisms*, for the dynamical complexity emerging from the composition of symmetric behaviors attracting the space to different invariant regions, as well as for composed computational models.

$$
\begin{array}{ccc}
S_i & \xrightarrow{\ \text{dyn.}\ } & \text{fixpoint} \\
\cup \downarrow & & \downarrow \text{union-inv.} \\
\cup_i S_i & \xrightarrow[\ \text{dyn.}\ ]{} & \text{complexity}
\end{array}
$$

Clearly, abstraction and complexity should always be treated on equal footing with composition to further develop the analysis of dynamical systems. We could systematically build relations between concrete and abstract systems, and between composed systems and their components.

Does composition preserve abstraction? Is composition monotonic w.r.t. abstraction? Given a concrete composition $\star$ and an appropriate abstraction function $\mathcal{Z}$, is there an abstract composition operator $\star'$ such that $\forall i, f_i \Subset g_i$ implies $\star_i f_i \Subset \star'_i g_i$?

If, moreover, the concrete global property results from the composition of individual properties, viz. $G_c(\star_i f_i) = \diamond_i I_c(f_i)$, does a corresponding equivalence hold at the abstract level, viz. $G_a(\star'_i g_i) = \diamond'_i I_a(g_i)$? In other words, and this will be our last question, does this last diagram commute?

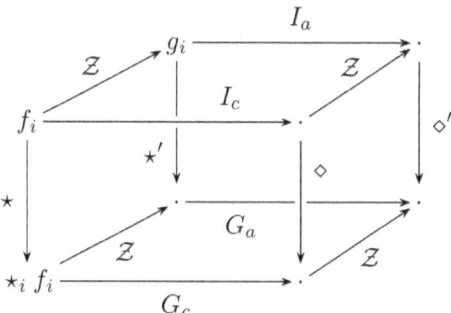

10.4 Coda: Compositional Complexity Revisited

The old Chinese tradition speaks of Yin and Yang as a duality attraction-opposition generating everything: male and female, Evil and Good. Our approach is oriented to the analysis of systems based on their structural construction and composition. In some sense, it reproduces what this Ancient Chinese philosophy had imagined thousands of years ago: attraction and opposition generate complexity, that is, order in disorder...

Bibliography

1. M. Abadi and L. Lamport. Composing specifications. *ACM Trans. on Programming Languages and Systems*, 15:73–132, 1993.
2. S. Abramsky, D. M. Gabbay, and T. S. E. Maibaum, editors. *Handbook of Logic in Computer Science*, volume 1 (Background: Mathematical Structures). Oxford Science Publications, 1992.
3. S. Abramsky, D. M. Gabbay, and T. S. E. Maibaum, editors. *Handbook of Logic in Computer Science*, volume 4 (Semantic Modelling). Oxford Science Publications, 1995.
4. A. Adamatzky. *Identification of Cellular Automata*. Taylor & Francis Ltd, 1994.
5. J. Adamek and J. Reiterman. Banach's fixed-point theorem as a base for data-type equations. *Applied Categorical Structures*, 2:77–90, 1994.
6. Y. Afek and Y. Matias. Elections in anonymous networks. *Information and Computation*, 113:312–330, 1994.
7. Z. Agur. Fixed points of majority rule cellular automata with application to plasticity and precision of the immune system. *Complex Systems*, 2:351–357, 1988.
8. Z. Agur, A. S. Fraenkel, and S. T. Klein. The number of fixed points of the majority rule. *Discrete Mathematics*, 70:295–302, 1988.
9. E. Akin. *The General Topology of Dynamical Systems*. American Mathematical Society, 1993.
10. J.-P. Allouche. The number of factors in a paperfolding sequence. *Bull. Austr. Math. Soc.*, 46:23–32, 1992.
11. J.-P. Allouche. Complexity of infinite sequences and the Ising transducer. In N. Boccara, E. Goles, S. Martinez, and P. Picco, editors, *Cellular Automata and Cooperative Systems*, volume 396 of *NATO ASI Ser. C: Math. Phys. Sci.*, pages 1–9. Kluwer Academic Publishers, 1993.
12. J.-P. Allouche. Sur la complexité des suites infinies. *Bull. Belg. Math. Soc.*, 1:133–143, 1994.
13. J.-P. Allouche and R. Bacher. Toeplitz sequences, paperfoldings, towers of hanoi and progression free sequences of integers. *Ens. Math.*, 38:315–327, 1992.
14. J.-P. Allouche and M. Bousquet-Mélou. Canonical positions for the factors in paperfolding sequences. *Theoretical Computer Science*, 129:263–278, 1994.
15. B. Alpern and F. B. Schneider. Defining liveness. *Information Processing Letters*, 21:181–185, 1985.
16. R. Alur, C. Courcoubetis, N. Halbwachs, T. A. Henzinger, P. H. Ho, X. Nicollin, A. Olivero, J. Sifakis, and S. Yovine. The algorithmic analysis of hybrid systems. *Theoretical Computer Science*, 138(1):3–34, 1995.
17. T. Amisaki, Y. Tsujino, and N. Tokura. Formal derivation of a probabilistic self-stabilizing program: Leader election on a uniform tree. In *Proc. of the 2nd*

Workshop on Self-Stabilizing Systems, Las Vegas. Dept. Computer Science, U. of Nevada, Las Vegas, 1995.

18. S. I. Andersson, Å.E. Andersson, and U. Ottoson, editors. *Theory and Control of Dynamical Systems, Applications to Systems in Biology, Proceedings of the 1991 Summer University of Southern Stockholm, Huddinge.* World Scientific, 1991.

19. A. Arnold. *Systèmes de Transitions Finis et Sémantique des Processus Communicants.* Masson, 1992.

20. A. Arnold. Hypertransition systems. In K. W. Wagner P. Enjalbert, E. W. Mayr, editor, *Proc. of STACS 94, Caen, France, Feb. 1994*, volume 775 of *Lecture Notes in Computer Science*, pages 327–338. Springer-Verlag, 1994.

21. V. I. Arnold and A. Avez. *Ergodic Problems of Classical Mechanics.* Addison-Wesley, 1968.

22. W. B. Arthur. Complexity in economic and financial markets. *Complexity*, 1(1):20–25, 1995.

23. E. Asarin and O. Maler. On some relations between dynamical systems and transition systems. In *Proceedings of ICALP'94*, volume 820 of *Lecture Notes in Computer Science*. Springer-Verlag, 1994.

24. D. Assaf IV and S. Gadbois. Definition of chaos (letter). *The American Mathematical Monthly*, 99(9):865, 1992.

25. G. Baier and M. Klein, editors. *A Chaotic Hierarchy.* World Scientific, 1991.

26. J. Banks, J. Brooks, G. Cairns, G. Davis, and P. Stacey. On Devaney's definition of chaos. *The American Mathematics Monthly*, 99(4):332–334, 1992.

27. A. Barbé, F. von Haeseler, H.-O. Peitgen, and G. Skordev. Coarse-graining invariant patterns of one-dimensional two-state linear cellular automata. *Int. J. Bifurcation and Chaos*, 5(6), 1995.

28. M. F. Barnsley. *Fractals Everywhere.* Academic Press, 1988.

29. M. F. Barnsley and L. P. Hurd. *Fractal Image Compression.* AK Peters, Ltd., 1992.

30. R. Bartlett. *Discrete Computation in the Continuum.* PhD thesis, Department of Mathematical Sciences, Memphis State University, 1994.

31. V. Bauchau. Les automates cellulaires : une approche informatique de la complexité du vivant. *Athena*, (73), 1991.

32. F. L. Bauer and H. Wössner. *Algorithmic Language and Program Development.* Springer-Verlag, 1982.

33. C. Beck and F. Schlögl. *Thermodynamics of Chaotic Systems.* Cambridge University Press, 1993.

34. T. Bedford, M. Keane, and C. Series, editors. *Ergodic Theory, Symbolic Dynamics and Hyperbolic Spaces.* Oxford Science Publications, 1991.

35. E. D. Beinhocker. Strategy at the edge of chaos. *The Mc Kinsey Quarterly*, (1):24–39, 1997.

36. C. Bercoff. A family of tag systems for paperfolding sequences. In E. W. Mayr and C. Puech, editors, *Proc. of STACS 95, Munich, Germany*, volume 900 of *Lecture Notes in Computer Science*, pages 303–312. Springer-Verlag, 1995.

37. H. M. Bizek. *Mathematics of the Rubik's Cube Design.* Dorrance Publishing Co., 1997.

38. F. Blayot and M. Verleysen. *Réseaux de neurones artificiels.* Que Sais-Je ? PUF, 1996.

39. M. M. Bonsangue and J. N. Kok. Relating multifunctions and predicate transformers through closure operators. In M. Hagiya and J. C. Mitchell, editors, *Proceedings of TACS 94, Sendai, Japan*, volume 789 of *Lecture Notes in Computer Science*, pages 822–843. Springer-Verlag, 1994.

40. F. Borceux and D. Dejean. Cauchy completion in category theory. *Cahiers de Topologie et Géométrie Différentielle Catégoriques*, XXVII(2):133–146, 1986.

41. G. Braga, G. Cattaneo, P. Flocchini, and G. Mauri. Complex chaotic behavior of a class of subshift cellular automata. *Complex Systems*, 7:269–296, 1993.

42. G. Braga, G. Cattaneo, P. Flocchini, and C. Quaranta Vogliotti. Pattern growth in elementary cellular automata. *Theoretical Computer Science*, 145:1–26, 1995.

43. C. Calude. *Information and Randomness. An Algorithmic Perspective*. Springer-Verlag, 1994.

44. J. L. Casti. *Reality Rules: Picturing the World in Mathematics (2 vol.)*. John Wiley & Sons, 1992.

45. J. L. Casti. *Reality Rules: Picturing the World in Mathematics, The Frontier*, volume 2. John Wiley & Sons, 1992.

46. J. L. Casti. *Reality Rules: Picturing the World in Mathematics, The Fundamentals*, volume 1. John Wiley & Sons, 1992.

47. G. Cattaneo, M. Finelli, and L. Margara. Topological chaos for elementary cellular automata. In *Proceedings of the 3rd Italian Conference onn Algorithms and Complexity*, 1997.

48. G. Cattaneo, P. Flocchini, G. Mauri, and N. Santoro. Chaos and subshift rules in neural networks and cellular automata. In *Proc. International Symposium on Nonlinear Theory and its Applications, Hawaii*, volume 4, pages 1153–1156. IEICE, 1993.

49. G. Cattaneo, P. Flocchini, G. Mauri, and N. Santoro. A new classification of cellular automata and their algebraic properties. In *Proc. International Symposium on Nonlinear Theory and its Applications, Hawaii*, volume 1, pages 223–226. IEICE, 1993.

50. G. Cattaneo, P. Flocchini, G. Mauri, and N. Santoro. Cellular automata in fuzzy backgrounds. *Physica D*, 105(1-3):105–120, 1997.

51. G. Cattaneo, E. Formenti, G. Manzini, and L. Margara. On ergodic linear cellular automata over Z_m. In R. Reischuk and M. Morvan, editors, *Proceeding of the 14th Symposium on Theoretical Aspects of Computer Science, Lubeck, March 97*, volume 1200 of *Lecture Notes in Computer Science*, pages 427–438. Springer-Verlag, 1997.

52. G. Cattaneo, E. Formenti, L. Margara, and J. Mazoyer. A shift-invariant metric on S^Z inducing a non-trivial topology. 1997.

53. G. Cattaneo, R. Nani, and G. Braga. A tool for the analysis of cellular automata. In J. Mazoyer, editor, *Workshop on Cellular Automata, 25-26 November 1991, Lyon, France*, pages 7–12. ESPRIT BRA WG 3166, ASMICS, 1991.

54. K. Cattell and J. C. Muzio. Analysis of one-dimensional linear hybrid cellular automata over GF(q). *IEEE Transactions on Computers*, 45:782–792, 1996.

55. K. Cattell and J. C. Muzio. Synthesis of one-dimensional linear hybrid cellular automata. *IEEE Transactions on Computer-Aided Design*, 15:325–335, 1996.

56. G. J. Chaitin. *Algorithmic Information Theory*. Cambridge University Press, 1990.

57. G. J. Chaitin. *Information, Randomness and Incompleteness: Papers on Algorithmic Information Theory*. World Scientific, 2nd edition, 1990.

58. G. J. Chaitin. Randomness and complexity in pure mathematics. *Int. Jnl. of Bifurcation and Chaos*, 4(1), 1994.

59. S. Chakraborty, D. Roy Chowdhury, and P. Pal Chaudhuri. Theory and applications of nongroup CA for synthesis of easily testable finite state machines. *IEEE Transactions on Computers*, 45(8):769–781, 1996.

60. K. M. Chandy. Mathematics of program construction applied to analog neural networks. In J. van de Snepscheut, editor, *Mathematics of Program Construction*, volume 375 of *Lecture Notes in Computer Science*, pages 21–35. Springer-Verlag, 1989.

61. K. M. Chandy and J. Misra. *Parallel Program Design, A Foundation*. Addison-Wesley, 1989.

62. K. M. Chandy and B. A. Sanders. Predicate transformers for reasoning about concurrent computation. *Science of Computer Programming*, 24:129–148, 1995.

63. H. Chaté and P. Manneville. Criticality in cellular automata. *Physica D*, 45:122–135, 1990.

64. C.-T. Chou. Simple proof techniques for property preservation via simulation. *Information Processing Letters*, 60:129–134, 1996.

65. P. Clérin, S. Grnac, P. Kuntz, and D. Snyers. Une nouvelle heuristique stochastique pour le routage dynamique, pour l'optimisation et la décision. In *Trois. Renc. Math. pour l'Optimisation et la Décision, Brest*, 1995.

66. P. Collet and J.-P. Eckmann. *Iterated Maps on the Interval as Dynamical Systems*. Birkhauser, 1980.

67. P. Collette. An explanatory presentation of composition rules for assumption-commitment specifications. *Information Processing Letters*, 50:31–35, 1994.

68. M. Cosnard, M. Garzon, and P. Koiran. Computability properties of low-dimensional dynamical systems. In P. Enjalbert, A. Finkel, and K. W. Wagner, editors, *Proc. of STACS 93*, volume 665 of *Lecture Notes in Computer Science*, pages 365–373. Springer-Verlag, 1993.

69. P. J. Courtois. On time and space decomposition of complex structures. *CACM*, 28(6):590–603, 1985.

70. P. Cousot. *Méthodes Itératives de Construction et d'Approximation de Points Fixes d'Operateurs Monotones sur un Treillis, Analyse Sémantique des Programmes*. PhD thesis, Université Scientifique et Médicale de Grenoble, 1978.

71. P. Cousot and R. Cousot. Constructive versions of Tarski's fixed point theorem. *Pacific J. Math.*, 82(1):43–57, 1979.

72. J. P. Crutchfield. The calculi of emergence: Computation, dynamics, and induction. *Physica D*, 75:11–54, 1994.

73. K. Culik II and S. Dube. L-systems and mutually recursive function systems. *Acta Informatica*, 30(3):279–302, 1993.

74. K. Culik II, L. P. Hurd, and S. Yu. Computation theoretic aspects of cellular automata. *Physica D*, 45:357–378, 1990.

75. K. Culik II, L. P. Hurd, and S. Yu. Formal languages and global cellular automaton behavior. *Physica D*, 45:396–403, 1990.

76. O.-J. Dahl. Discrete event simulation languages. In F. Genuys, editor, *Programming Lannguages*, pages 349–395. Academic Press, 1968.

77. J. Dassow and J. Kelemen. Cooperating/distributed grammar systems: a link between formal languages and artificial intelligence. *Bulletin of the EATCS*, (45):131–145, 1991.

78. B. A. Davey and H. A. Priestley. *Introduction to Lattices and Order*. Cambridge University Press, 1990.

79. J. Davies. *Specification and Proof in Real-Time CSP*. Cambridge University Press, 1993.

80. C. Davis and D. E. Knuth. Number representations and dragon curves – I & II. *Journal of Recreational Mathematics*, 3:61–81, 133–149, 1970.

81. J. W. de Bakker, W. P. de Roever, and G. Rozenberg, editors. *A Decade of Concurrency, Reflections and Perspectives*, volume 803 of *Lecture Notes in Computer Science*. Springer-Verlag, 1994.

82. P. De Grauwe, H. Dewachter, and M. Embrechts. *Exchange Rate Theory – Chaotic Models of Foreign Exchange Markets.* Blackwell Publishers, 1993.
83. M. Dekking, M. Mendès France, and A. van der Poorten. Folds! *Math. Intell.*, 4:130–138; 173–181; 190–195, 1982.
84. J.-P. Delahaye. *Information, Complexité et Hasard.* Hermès, 1994.
85. J.-P. Delahaye. Complexité des objets composés. Technical report, U.S.T. Lille, URA CNRS 369, 1995.
86. N. Dershowitz. Termination of rewriting. *J. Symbolic Computation*, 3:69–116, 1987.
87. D. Deutsch. Quantum theory, the Church-Turing principle and the universal quantum computer. In *Proc. Royal Soc. London*, volume A400, pages 97–117, 1985.
88. R. L. Devaney. *An Introduction to Chaotic Dynamical Systems.* Addison-Wesley, 2nd edition, 1989.
89. V. Diekert and G. Rozenberg, editors. *The Book of Traces.* World Scientific, 1995.
90. E. W. Dijkstra. Self-stabilizing systems in spite of distributed control. *Communications of the ACM*, 17(11):643–644, 1974.
91. E. W. Dijkstra. *A Discipline of Programming.* Prentice Hall, 1976.
92. E. W. Dijkstra. Edw675: The equivalence of bounded nondeterminacy and continuity. In *Selected Writings on Computing: a Personal Perspective*, pages 358–359. Springer-Verlag, 1982.
93. E. W. Dijkstra and C. S. Scholten. *Predicate Calculus and Program Semantics.* Springer-Verlag, 1990.
94. W. Dörfler. The cartesian composition of automata. *Math. Systems Theory*, 11:239–257, 1978.
95. M. Dorigo. The ant system: Optimization by a colony of cooperating agents. *IEEE Transactions on Systems, Man, and Cybernetics (B)*, 26(1):1–13, 1996.
96. J. Dugundji. *Topology.* Wm.C. Brown Publishers, 2nd edition, 1989.
97. A. Edalat. Dynamical systems, measures, and fractals via domain theory. *Information and Computation*, 120:32–48, 1995.
98. E. A. Emerson. Temporal and modal logic. In J. Van Leeuwen, editor, *Handbook of Theoretical Computer Science*, volume B, pages 995–1072. Elsevier, 1990.
99. E. A. Emerson and E. M. Clarke. Using branching time temporal logic to synthesize synchronization skeletons. *Science of Computer Programming*, 2:241–266, 1982.
100. Y. Fisher, editor. *Fractal Image Compression.* Springer-Verlag, 1995.
101. P. Flocchini and F. Geurts. Searching for chaos in cellular automata: Compositional approach. In R. J. Stonier and X. H. Yu, editors, *Complex Systems, Mechanism of Adaptation*, pages 329–336. IOS Press, 1994.
102. P. Flocchini and F. Geurts. Searching for chaos in cellular automata: New tools for classification. In R. J. Stonier and X. H. Yu, editors, *Complex Systems, Mechanism of Adaptation*, pages 321–328. IOS Press, 1994.
103. P. Flocchini, F. Geurts, A. Mingarelli, and N. Santoro. Convergence and aperiodicity in continuous cellular automata. Technical Report RR 97-01, Département d'Ingénierie Informatique, Université catholique de Louvain, 1997.
104. P. Flocchini, F. Geurts, and N. Santoro. Dynamics and algebraic properties of moss reinforced shifts. In *Proc. International Symposium on Nonlinear Theory and its Applications, Las Vegas*, volume 2, pages 1149–1152. IEICE, 1995.

105. P. Flocchini, F. Geurts, and N. Santoro. Compositional experimental analysis of cellular automata: Attraction properties and logic disjunction. Technical Report TR-96-31, School of Computer Science, Carleton University, 1996. http://www.scs.carleton.ca/scs/tech_reports/1996/list.html; also: RR 96-10, Département d'Informatique, Université catholique de Louvain.

106. P. Flocchini, F. Geurts, and N. Santoro. CA-like error propagation in fuzzy CA. *Parallel Computing*, 23(11):1673–1682, 1997.

107. P. Flocchini and N. Santoro. The chaotic evolution of information in the interaction between knowledge and uncertainty. In R. J. Stonier and X. H. Yu, editors, *Complex Systems, Mechanism of Adaptation*, pages 337–343. IOS Press, 1994.

108. P. Flocchini and N. Santoro. Topological constraints for sense of direction. In L. M. Kirousis and E. Kranakis, editors, *Proc. of the 2nd Colloquium on Structural Information and Communication Complexity*, pages 27–38. Carleton University Press, 1996.

109. R. W. Floyd. Assigning meanings to programs. In *Proc. Symp. Applied Mathematics*, volume 19, pages 19–32. Amer. Math. Soc., 1967.

110. N. Francez. *Fairness*. Springer-Verlag, 1986.

111. S. Franklin and M. Garzon. *On Stability and Solvability (Or, When Does a Neural Network Solve a Problem?)*, volume 2, pages 71–83. Kluwer Academic Publishers, 1992.

112. P. H. H. Gardiner, C. E. Martin, and O. de Moor. An algebraic construction of predicate transformers. *Science of Computer Programming*, 22:21–44, 1994.

113. M. Garzon. *Models of Massive Parallelism. Analysis of Cellular Automata and Neural Networks*. Springer-Verlag, 1995.

114. M. Garzon and S. Franklin. Neural computability II. In *Proc. 3rd Int. Joint Conf. on Neural Networks*, 1989.

115. R. J. Gaylord and K. Nishidate. *Modeling Nature: Cellular Automata Simulations with Mathematica*. Springer-Verlag, 1996.

116. F. Geurts and V. Lombart. Etude des systèmes de transitions discrets. Diploma Thesis, Unité d'Informatique, Université catholique de Louvain, 1992.

117. A. Ginzburg. *Algebraic Theory of Automata*. Academic Press, 1968.

118. J. Gleick. *Chaos*. The Viking Press, 1987.

119. E. Goles. Lyapunov operators to study the convergence of extremal automata. *Theoretical Computer Science*, 125:329–337, 1994.

120. E. Goles, F. Fogelman-Soulie, and D. Pellegrin. Decreasing energy fonction as a tool for studying threshold networks. *Discrete Applied Mathematics*, 12(3):261–277, 1985.

121. E. Goles and S. Martinez. *Neural and Automata Networks, Dynamical Behavior and Applications*. Kluwer Academic Publishers, 1990.

122. E. Goles and J. Olivos. Periodic behavior of generalized threshold functions. *Discrete Mathematics*, 30:187–189, 1980.

123. M. G. Gouda. The triumph and tribulation of system stabilization. In J. M. Hélary and M. Raynal, editors, *Proc. of the 9th International Workshop on Distributed Algorithms*, volume 972 of *Lecture Notes in Computer Science*, pages 1–18. Springer-Verlag, 1995.

124. A. Granville. On a paper by Agur, Fraenkel and Klein. *Discrete Mathematics*, 94:147–151, 1991.

125. P. Grassberger. Chaos and diffusion in deterministic cellular automata. *Physica D*, 10:52, 1984.

126. G. Grössing and A. Zeilinger. Quantum cellular automata (+ a corrigendum). *Complex Systems*, 2(2/5):197–209/611–623, 1988.

127. R. L. Grossman and M. Sweedler. Hybrid systems and quantum automata: Preliminary announcement. In P. Antsaklis, W. Kohn, A. Nerode, and S. Sastry, editors, *Hybrid Systems II*, volume 999 of *Lecture Notes in Computer Science*, pages 191–201. Springer-Verlag, 1995.

128. P. Guan. Cellular automaton public-key cryptosystem. *Complex Systems*, 1:51–57, 1987.

129. J. Guckenheimer. A robust hybrid stabilization strategy for equilibria. *IEEE Trans. Autom. Control*, 40(2):321–326, 1995.

130. P. Guerreiro. A relational model for non-deterministic programs and predicate transformers. In B. Robinet, editor, *Proceedings of the International Symposium on Programming*, volume 83 of *Lecture Notes in Computer Science*, pages 136–146. Springer-Verlag, 1980.

131. R. Gupta, S. A. Smolka, and S. Bhaskar. On randomization in sequential and distributed algorithms. *ACM Computing Surveys*, 26(1):7–86, 1994.

132. H. Gutowitz, editor. *Cellular Automata, Theory and Experiment*. MIT Press/North-Holland, 1991.

133. H. Gutowitz and C. Langton. Mean field theory of the edge of chaos. In F. Moran, A. Moreno, J. J. Merelo, and P. Chacon, editors, *Advances in Artificial Life, 3rd European Conference on Artificial Life, Granada*, volume 929 of *LNAI*, pages 52–64. Springer-Verlag, 1995.

134. H. A. Gutowitz. A hierachical classification of cellular automata. *Physica D*, 45:136–156, 1990.

135. H. A. Gutowitz. Cryptography with dynamical systems. In N. Boccara, E. Goles, S. Martinez, and P. Picco, editors, *Cellular Automata and Cooperative Systems*, volume 396 of *NATO ASI Ser. C: Math. Phys. Sci.*, pages 237–274. Kluwer Academic Publishers, 1993.

136. H. A. Gutowitz. Cellular automata and the sciences of complexity I & II. *Complexity*, 1(5/6):16–22/29–25, 1995/1996.

137. H. A. Gutowitz, J. D. Victor, and B. W. Knight. Local structure theory for cellular automata. *Physica D*, 28:18–48, 1987.

138. B. L. Hao, editor. *Chaos (Reprinted Papers)*. World Scientific, 1984.

139. B. L. Hao. *Elementary Symbolic Dynamics and Chaos in Dissipative Systems*. World Scientific, 1989.

140. M. Hata. On the structure of self-similar sets. *Japan J. Appl. Math.*, 2:381–414, 1985.

141. F. Hausdorff. *Set Theory*. Chelsea, 1962.

142. S. Haykin. *Neural Networks. A Comprehensive Foundation*. IEEE Computer Society Press, 1994.

143. G. A. Hedlund. Endomorphisms and automorphisms of the shift dynamical system. *Mathematical Systems Theory*, 3:320–375, 1969.

144. H. A. J. M. Heijmans and C. Ronse. The algebraic basis of mathematical morphology – part I: Dilatations and erosions. *Computer Vision, Graphics and Image Processing*, 50:245–295, 1990.

145. M. Hénon. A two-dimensional mapping with a strange attractor. *Commun. Math. Phys.*, 50:69–77, 1976.

146. R. Herken, editor. *The Universal Turing Machine, A Half-Century Survey*. Oxford University Press, 1988.

147. T. Herman. Probabilistic self-stabilization. *Information Processing Letters*, 35:63–67, 1990.

148. A. V. M. Herz and C. M. Marcus. Distributed dynamics in neural networks. *Physical Review E*, 47(3):2155–2161, 1993.

149. D. Hilbert and P. Bernays. *Grundlagen der Mathematik*, volume 2. Springer-Verlag, 1939.

150. C. A. R. Hoare. Some properties of predicate transformers. *Journal of the ACM*, 25(3):461–480, 1978.

151. C. A. R. Hoare. *Communicating Sequential Processes*. Prentice Hall, 1985.

152. C. A. R. Hoare and J. F. He. The weakest prespecification. *Information Processing Letters*, 24:127–132, 1987.

153. J. Holland. Les algorithmes génétiques. *Pour la Science*, (179):44–51, 1992.

154. R. A. Holmgren. *A First Course in Discrete Dynamical Systems*. Universitext. Springer-Verlag, 1994.

155. J. E. Hopcroft and J. D. Ullmann. *Introduction to Automata Theory, Languages, and Computation*. Addison-Wesley, 1990.

156. J. J. Hopfield. Neural networks and physical systems with emergent collective computational abilities. *Proc. of the National Academy of Sciences*, 79:2554–2558, 1982.

157. J. J. Hopfield. Neurons with graded response have collective computational properties like those of two-state neurons. *Proc. of the National Academy of Sciences*, 81:3088–3092, 1984.

158. B. A. Huberman and T. Hogg. The emergence of computational ecologies. In L. Nader and D. Stein, editors, *1992 Lectures in Complex Systems*, volume V of *SFI Studies in the Sci. of Complexity*. Addison-Wesley, 1993.

159. J. E. Hutchinson. Fractals and self similarity. *Indiana University Mathematics Journal*, 30(5):713–747, 1981.

160. J. E. Hutchinson. Fractals: a mathematical framework. In R. J. Stonier and X. H. Yu, editors, *Complex Systems: Mechanism of Adaptation*, pages 271–281. IOS Press, 1994.

161. E. Jen. Aperiodicity in one-dimensional cellular automata. *Physica D*, 45:3–18, 1990.

162. C. Jones. *Probabilistic Nondeterminism*. PhD thesis, ECS, LFCS, U. Edinburgh, 1990.

163. N. D. Jones and F. Nielson. Abstract interpretation: a semantics-based tool for program analysis. In S. Abramsky, D. M. Gabbay, and T. S. E. Maibaum, editors, *Handbook of Logic in Computer Science*, volume 4 Semantic Modelling, chapter 5, pages 527–636. Oxford Science Publications, 1995.

164. P. Kůrka. Simulation in dynamical systems and Turing machines. Technical report, Department of Mathematical Logic and Philosophy of Mathematics, Charles U., Prague, 1992.

165. P. Kůrka. Universal computation in dynamical systems. Technical report, Department of Mathematical Logic and Philosophy of Mathematics, Charles U., Prague, 1992.

166. P. Kůrka. Dynamical systems and factors of finite automata. Technical report, Department of Mathematical Logic and Philosophy of Mathematics, Charles U., Prague, 1993.

167. P. Kůrka. One-dimensional dynamics and factors of finite automata. *Acta Univ. Carolinae Math. Phys.*, 34(2):83–95, 1993.

168. P. Kůrka. A comparison of finite and cellular automata. In I. Prívara, B. Rovan, and P. Ružička, editors, *Mathematical Foundations of Computer Science*, volume 841 of *Lecture Notes in Computer Science*, pages 483–493. Springer-Verlag, 1994.

169. P. Kůrka. Regular unimodal systems and factors of finite automata. *Theoretical Computer Science*, 133:49–64, 1994.

170. P. Kůrka. Simplicity criteria for dynamical systems. In S. I. Andersson, editor, *Analysis of Dynamical and Cognitive Systems*, volume 888 of *Lecture Notes in Computer Science*, pages 189–225. Springer-Verlag, 1995.

171. P. Kůrka. Languages, equicontinuity and attractors in linear cellular automata. *Ergodic Theory and Dynamical Systems*, 16:1–17, 1996.

172. P. Kůrka. On topological dynamics of Turing machines. *Theoretical Computer Science*, 174:203–216, 1997.

173. Y. Kamp and M. Hasler. *Réseaux de neurones récursifs pour mémoires associatives*. Presses Polytechniques et Universitaires Romandes, 1990.

174. K. Kaneko. Attractors, basin structures and information processing in cellular automata. In S. Wolfram, editor, *Theory and Applications of Cellular Automata*, pages 367–399. World Scientific, 1986.

175. K. Kaneko. Homeochaos: Dynamic stability of a symbiotic network with population dynamics and evolving mutation rates. *Physica D*, 56:406–429, 1992.

176. K. Kaneko. *Theory and Application of Coupled Map Lattices*. John Wiley & Sons, 1993.

177. K. Kaneko. Chaos as a source of complexity and diversity in evolution. *Artificial Life*, 1(1/2):163–177, 1993–94.

178. D. Kaplan and L. Glass. *Understanding Nonlinear Dynamics*. Springer-Verlag, 1995.

179. E. Kindler. Invariants, compositionality and substitution. *Acta Informatica*, 32(4):299–312, 1995.

180. B. G. Klein. Homomorphisms of symbolic dynamical systems. *Math. Systems Theory*, 6:107–122, 1972.

181. C. Knudsen. Chaos without nonperiodicity. *American Mathematical Monthly*, pages 563–565, June-July 1994.

182. P. Koiran. *Puissance de Calcul des Réseaux de Neurones Artificiels*. PhD thesis, Laboratoire de l'Informatique du Parallélisme, E.N.S.Lyon, 1993.

183. P. Koiran and C. Moore. Closed-form analytic maps in one and two dimensions can simulate Turing machines. *Submitted to Theoretical Computer Science*, 1996. Available from http://www.santafe.edu/~moore.

184. J. R. Koza. *Genetic Programming: On the Programming of Computers by Means of Natural Selection*. MIT Press, 1992.

185. J. R. Koza. *Genetic Programming II: Automatic Discovery of Reusable Programs*. MIT Press, 1994.

186. M. A. Krasnoselskii, A. V. Lusnikov, and A. V. Pokrovskii. Stable fixed points of monotone operators. *Russian Acad. Sci. Dokl. Math.*, 47(3):427–430, 1993.

187. R. Kumar, V. Garg, and S. I. Marcus. Predicates and predicate transformers for supervisory control of discrete event dynamical systems. *IEEE Trans. Autom. Control*, 38(2):232–247, 1993.

188. P. Kuntz and D. Snyers. Emergent colonization and graph partitioning. In *Proc. 3rd Int. Conf. on Simulation of Adaptive Behavior*, pages 494–500. MIT Press, 1994.

189. M. Z. Kwiatkowska. A metric for traces. *Information Processing Letters*, 35:129–135, 1990.

190. M. Z. Kwiatkowska. On topological characterization of behavioral properties. In G. M. Reed, A. W. Roscoe, and R. F. Wachter, editors, *Topology and Category Theory in Computer Science*, Oxford Science Publications, chapter 6, pages 153–177. Clarendon Press, 1991.

191. O. Ladyzhenskaya. *Attractors for Semigroups and Evolution Equations*. Cambridge University Press, 1991.

192. L. Lamport. "sometime" is sometimes "not never". on the temporal logic of programs. In *Proc. 7th Annual ACM Symp. Princ. Prog. Lang., Las Vegas*, pages 174–185. ACM, 1980.

193. L. Lamport. A simple approach to specifying concurrent systems. *CACM*, 32(1):32–45, 1989.

194. L. Lamport. *win* and *sin*: Predicate transformers for concurrency. *ACM Trans. on Prog. Lang. and Syst.*, 12(3), 1990.

195. C. G. Langton. Computation at the edge of chaos: Phase transitions and emergent computation. *Physica D*, 42:12–37, 1990.

196. A. Lasota and M. C. Mackey. *Chaos, Fractals, and Noise, Stochastic Aspects of Dynamics.* Springer-Verlag, 2nd edition, 1994.

197. J. L. Lassez, V. L. Nguyen, and E. A. Sonenberg. Fixed point theorems and semantics: a folk tale. *Information Processing Letters*, 14(3):112–116, 1982.

198. P. Layzell and P. Kuntz. A new stochastic approach to find clusters in vertex set of large graphs with applications to partitioning in VLSI technology. Technical report, Telecom Bretagne, 1995.

199. M. Li and P. Vitànyi. *An Introduction to Kolmogorov Complexity and its Applications.* Springer-Verlag, 1993.

200. W. Li and N. H. Packard. The structure of the elementary cellular automata rule space. *Complex Systems*, 4:281–297, 1990.

201. W. Li, N. H. Packard, and C. G. Langton. Transition phenomena in cellular automata rule space. *Physica D*, 45:77–94, 1990.

202. D. Lind and B. Marcus. *An Introduction to Symbolic Dynamics and Coding.* Cambridge University Press, 1995.

203. K. Lindgren. Entropy and correlations in discrete dynamical systems. In J. L. Casti and A. Karlqvist, editors, *Beyond Belief: Randomness, Predication and Explanation in Science*, chapter 5, pages 88–109. CRC Press, 1991.

204. K. Lindgren and M. G. Nordahl. Universal computation in simple one-dimensional cellular automata. *Complex Systems*, 4:299–318, 1990.

205. B. Litow and P. Dumas. Additive cellular automata and algebraic series. *Theoretical Computer Science*, 119:345–354, 1993.

206. C. Loiseaux, S. Graf, J. Sifakis, A. Bouajjani, and S. Bensalem. Property preserving abstractions for the verification of concurrent systems. *Formal Methods in System Design*, 6:11–44, 1995.

207. E. N. Lorentz. Deterministic nonperiodic flow. *Journal of the Atmospheric Sciences*, 20:130–141, 1963.

208. B. B. Mandelbrot. *The Fractal Geometry of Nature.* Freeman, 1982.

209. B. B. Mandelbrot. *Fractal and Scaling in Finance: Discontinuity, Concentration, Risk.* Springer-Verlag, 1997.

210. Z. Manna and A. Pnueli. *The Temporal Logic of Reactive and Concurrent Systems: Specification.* Springer-Verlag, 1992.

211. C. M. Marcus, F. R. Waugh, and R. M. Westervelt. Associative memory in an analog iterated-map neural network. *Physical Review A*, 41(6):3355–3364, 1990.

212. C. M. Marcus and R. W. Westervelt. Dynamics of iterated-maps neural networks. *Physical Review A*, 40(1):501–504, 1989.

213. L. Margara. Cellular automata and non periodic orbits. *Complex Systems*, (to appear).

214. N. Margolus. Parallel quantum computation. In W. H. Zurek, editor, *Complexity, Entropy, and the Physics of Information*, volume VIII of *SFI Studies in the Sci. of Complexity*, pages 273–287. Addison-Wesley, 1990.

215. B. Martin. *Construction Modulaire d'Automates Cellulaires.* PhD thesis, Laboratoire de l'Informatique du Parallélisme, E.N.S.Lyon; Université Claude Bernard Lyon 1, 1993.

216. A. Mazurkiewicz. Trace theory. In W. Brauer, W. Reisig, and G. Rozenberg, editors, *Petri Nets: Applications and Relationships to Other Models of Concurrency*, volume 255 of *Lecture Notes in Computer Science*, pages 279–324. Springer-Verlag, 1986.

217. J. L. McCauley. *Chaos, Dynamics and Fractals: An Algorithmic Approach to Deterministic Chaos.* Cambridge University Press, 1993.
218. M. Mendès France and J. O. Shallit. Wire bending. *Journal of Combinatorial Theory A*, 50:1–23, 1989.
219. M. Mendès France and A. J. van der Poorten. Arithmetic and analytic properties of paper folding sequences. *Bull. Austral. Math. Soc.*, 24:123–131, 1981.
220. D. A. Meyer. Towards the global: Complexity, topology and chaos in modelling, simulation and computation. 1997. Available from http://xxx.lanl.gov/abs/chao-dyn/9710005.
221. A. N. Michel and K. Wang. *Qualitative Theory of Dynamical Systems. The Role of Stability Preserving Mappings.* Marcel Dekker Inc., 1995.
222. R. Milner. *Communication and Concurrency.* Prentice Hall, 1989.
223. M. L. Minsky. *Perceptrons: An Introduction to Computational Geometry.* MIT Press, 3rd edition, 1988.
224. C. Moore. Unpredictability and undecidability in dynamical systems. *Physical Review Letters*, 64(20):2354–2357, 1990.
225. C. Moore. Generalized one-sided shifts and maps of the interval. *Nonlinearity*, 4:727–745, 1991.
226. C. Moore. Generalized shifts: Unpredictability and undecidability in dynamical systems. *Nonlinearity*, 4:199–230, 1991.
227. C. Moore. Smooth maps of the interval and the real line capable of universal computation. Technical Report 93-01-001, Santa Fe Institute, 1993. Available from http://www.santafe.edu/~moore.
228. C. Moore. Dynamical recognizers: Real-time language recognition by analog computers. *Submitted to Theoretical Computer Science*, 1996. Available from http://www.santafe.edu/~moore.
229. C. Moore. Recursion theory on the reals and continuous-time computation. *Theoretical Computer Science*, 162:23–44, 1996.
230. C. Moore and E. Aurell. Symbolic dynamics, transcendentality, and complexity at the transition to chaos. Technical report, Dept. Physics, Cornell, and Inst. Theor. Physics, Göteborg, 1990. Available from http://www.santafe.edu/~moore.
231. G. Moran. Parametrization for stationary patterns of the r-majority operators on 0–1 sequences. *Discrete Mathematics*, 132:175–195, 1994.
232. G. Moran. The r-majority vote action on 0–1 sequences. *Discrete Mathematics*, 132:145–174, 1994.
233. G. Moran. On the period-two-property of the majority operator in infinite graphs. *Transactions of the American Mathematical Society*, 347(5):1649–1667, 1995.
234. C. Morgan, A. McIver, and K. Seidel. Probabilistic predicate transformers. *TOPLAS*, 18(3):325–353, 1996.
235. C. Morgan and T. Vickers, editors. *On the Refinement Calculus.* Springer-Verlag, 2nd edition, 1994.
236. C. C. Morgan. Proof rules for probabilistic loops. In H. Jifeng, J. Cooke, and P. Wallis, editors, *Proceedings of the BCS-FACS 7th Refinement Workshop.* Springer-Verlag, 1996.
237. C. C. Morgan and A. McIver. Unifying wp and wlp. *Information Processing Letters*, 20(3):159–164, 1996.
238. S. Mullender, editor. *Distributed Systems.* Addison-Wesley, 2nd edition, 1993.
239. S. Nandi, B. K. Kar, and P. Pal Chaudhuri. Theory and application of cellular automata in cryptography. *IEEE Transactions on Computers*, 43(12):1346–1357, 1994.

240. A. Nayak, L. Pagli, and N. Santoro. Efficient construction of catastrophic patterns for vlsi reconfigurable arrays. *INTEGRATION, the VLSI Journal,* 15:133–150, 1993.

241. A. Nayak, L. Pagli, and N. Santoro. On testing for catastrophic faults in reconfigurable arrays with arbitrary link redundancy. *INTEGRATION, the VLSI Journal,* 20:327–342, 1996.

242. A. Nayak, N. Santoro, and R. Tan. Fault-tolerance of reconfigurable systolic arrays. In *Proc. of the 20th International Symposium on Fault-Tolerant Computing, Newcastle, UK,* pages 202–209. IEEE Computer Society, 1990.

243. G. L. Nemhauser and L. A. Wolsey. *Integer and Combinatorial Optimization.* John Wiley & Sons, 1988.

244. A. Nerode and W. Kohn. Models of hybrid systems: Automata, topologies, controllability, observability. In R. L. Grossman, A. Nerode, A. P. Ravn, and H. Rischel, editors, *Hybrid Systems,* volume 736 of *Lecture Notes in Computer Science,* pages 317–356. Springer-Verlag, 1993.

245. T. T. Nguyen. *Multi-Valued Function Theory for Computer Programming.* PhD thesis, Département de Mathématique, Université catholique de Louvain, 1988.

246. T. T. Nguyen. A relational model of demonic nondeterministic programs. *International Journal of Foundations of Computer Science,* 2(2):101–131, 1991.

247. E. Ohlebusch. On the modularity of termination of term rewriting systems. *Theoretical Computer Science,* 136:333–360, 1994.

248. D. S. Ornstein. *Ergodic Theory, Randomness, and Dynamical Systems.* Number 5 in Yale Math. Monographs. Yale University Press, 1974.

249. P. Pal Chaudhuri, D. Roy Chowdhury, S. Nandi, and S. Chattopadhyay. *Theory and Applications in Additive Cellular Automata.* IEEE Press, 1997.

250. J. Palis and F. Takens. *Hyperbolicity and Sensitive Chaotic Dynamics at Homoclinic Bifurcations.* Cambridge University Press, 1993.

251. F. Pasemann. Discrete dynamics of two neuron networks. Technical report, Institut für Theoretische Physik, T.U. Clausthal, Germany, 1993.

252. F. Pasemann. Dynamics of a single model neuron. *International Journal of Bifurcation and Chaos,* 2:271–278, 1993.

253. F. Pasemann. Characterization of periodic attractors in neural ring networks. *Neural Networks,* 8(3):421–429, 1995.

254. F. Pasemann. A simple chaotic neuron. *Physica D,* 104(2):205–211, 1997.

255. F. Pasemann and E. Nelle. Dynamical effects of coupling two neurons. Technical report, Institut für Theoretische Physik, T.U. Clausthal, Germany, 1993.

256. G. Păun. Splicing: A challenge for formal language theorists. *Bulletin of the EATCS,* (57):183–194, 1995.

257. J. Pedersen. Continuous transitions of cellular automata. *Complex Systems,* 4:653–665, 1990.

258. H.-O. Peitgen and P. H. Richter. *The Beauty of Fractals.* Springer-Verlag, 1986.

259. H.-O. Peitgen, G. Skordev, and F. von Haeseler. Global analysis of self-similarity features of cellular automata: Selected examples. *Physica D,* 86:64–80, 1995.

260. H.-O. Peitgen, G. Skordev, and F. von Haeseler. Multifractal decomposition of rescaled evolution sets of equivariant cellular automata. *Journal of Random and Comput. Dynamics,* 3:93–119, 1995.

261. D. Peleg. Local majority voting, small coalitions and controlling monopolies in graphs: A review. In N. Santoro and P. Spirakis, editors, *Structure, Information and Communication Complexity, 3rd Colloquium SIROCCO'96, Certosa di Pontignano, Italy,* pages 152–169. Carleton University Press, 1997.

262. G. L. Peterson. Myths about teh mutual exclusion problem. *Information Processing Letters*, 12(3):115–116, 1981.

263. M. Phipps. From local to global: the lesson of cellular automata. In D. DeAngelis and L. Gross, editors, *Individual-Based Approaches in Ecology: Concepts and Models*. Chapman and Hall, 1992.

264. S. Y. Pilyugin. *The Space of Dynamical Systems with the C^0-Topology*, volume 1571 of *Lecture Notes in Mathematics*. Springer-Verlag, 1994.

265. B. Plateau and K. Atif. Stochastic automata network for modeling parallel systems. *IEEE Trans. Soft. Eng.*, 17(10):1093–1108, 1991.

266. S. Poljak. Transformations on graphs and convexity. *Complex Systems*, 1:1021–1033, 1987.

267. S. Poljak and M. Sůra. On periodical behavior in societies with symmetric influences. *Combinatorica*, 1:119–121, 1983.

268. M. Queffélec. *Substitution Dynamical Systems – Spectral Analysis*, volume 1294 of *Lecture Notes in Mathematics*. Springer-Verlag, 1987.

269. P. Ramanathan, K. G. Shin, and R. W. Butler. Fault-tolerant clock synchronisation in distributed systems. *IEEE Trans. Computer*, pages 33–42, Oct. 1990.

270. D. A. Rand. Measuring and characterizing spatial patterns, dynamics and chaos in spatially extended dynamical systems and ecologies. *Phil. Trans. R. Soc. Lond. A*, 348:497–514, 1994.

271. J. R. Rao. Reasoning about probabilistic parallel programs. *ACM Trans. on Prog. Lang. and Syst.*, 16(3):798–842, 1994.

272. S. Rasmussen and C. L. Barrett. Elements of a theory of simulation. In F. Moran, A. Moreno, J. J. Merelo, and P. Chacon, editors, *Advances in Artificial Life, 3rd European Conference on Artificial Life, Granada*, volume 929 of *LNAI*, pages 515–529. Springer-Verlag, 1995.

273. D. Richardson. Tessellations with local transformations. *Journal of Computer and System Sciences*, 6:373–388, 1972.

274. F. Robert. *Discrete Iterations, A Metric Study*. Springer-Verlag, 1986.

275. C. Ronse and H. A. J. M. Heijmans. The algebraic basis of mathematical morphology – part II: Openings and closings. *Computer Vision, Graphics and Image Processing*, 54:74–97, 1991.

276. G. Schmidt and T. Ströhlein. Timetable constructio: An annotated bibliography. *The Computer Journal*, 23(4):307–316, 1980.

277. G. Schmidt and T. Ströhlein. *Relations and Graphs*. Springer-Verlag, 1993.

278. M. Schneider. Self-stabilization. *ACM Computing Surveys*, 25(1):45–67, 1993.

279. M. Schroeder. *Fractals, Chaos, Power Laws*. W. H. Freeman and Co., 1991.

280. J. Serra. *Image Analysis and Mathematical Morphology*. AcademicPress, 1982.

281. M. Serra, T. Slater, J. C. Muzio, and D. M. Miller. The analysis of one-dimensional linear cellular automata and their aliasing properties. *IEEE Transactions on Computer-Aided Design*, 9:767–778, 1990.

282. H. T. Siegelmann. Computation beyond the Turing limit. *Science*, 268:545–548, 1995.

283. H. T. Siegelmann and E. D. Sontag. Analog computation via neural networks. *Theoretical Computer Science*, 131:331–360, 1994.

284. J. Sifakis. A unified approach for studying the properties of transition systems. *Theoretical Computer Science*, 18:227–258, 1982.

285. Y. G. Sinai. *Topics in Ergodic Theory*. Number 44 in Math. Ser. Princeton University Press, 1994.

286. M. Sintzoff. Invariance and contraction by infinite iterations of relations. In J.-P. Banâtre and D. Le Metayer, editors, *Research Directions in High-Level Parallel Programming Languages*, volume 574 of *Lecture Notes in Computer Science*, pages 349–373. Springer-Verlag, 1992.

287. M. Sintzoff. Invariance and termination in structured dynamical systems. In *Proc. 1995 Intl Symp. on Nonlinear Theory and its Applications*, volume 1. IEICE Tokyo, 1995.

288. M. Sintzoff. Abstract verification of structured dynamical systems. In *Proc. of Hybrid Systems III*, Lecture Notes in Computer Science. Springer-Verlag, 1996.

289. M. Sintzoff and F. Geurts. Compositional analysis of dynamical systems using predicate transformers (summary). In *Proc. International Symposium on Nonlinear Theory and its Applications, Hawaii*, volume 4, pages 1323–1326. IEICE, 1993.

290. M. Sintzoff and F. Geurts. Analysis of dynamical systems using predicate transformers: Attraction and composition. In S. I. Andersson, editor, *Analysis of Dynamical and Cognitive Systems*, volume 888 of *Lecture Notes in Computer Science*, pages 227–260. Springer-Verlag, 1995.

291. M. Sipper. *Evolution of Parallel Cellular Machines – The Cellular Programming Approach*, volume 1194 of *Lecture Notes in Computer Science*. Spring-Verlag, 1997.

292. S. Smale. Diffeomorphisms with many periodic points. In S. S. Cairns, editor, *Differential and Combinatorial Topology*, pages 63–80. Princeton University Press, 1965.

293. S. Smale. Differential dynamical systems. *Bull. of the Amer. Math. Soc.*, 73:747–817, 1967.

294. D. R. Smith. Constructing specification morphisms. *Journal of Symbolic Computation*, 15(5–6):571–606, 1993.

295. D. R. Smith and M. R. Lowry. Algorithms theories and design tactics. *Science of Computer Programming*, 14(2–3):305–321, 1990.

296. A. R. Smith III. Simple computation-universal cellular spaces. *Journal of the ACM*, 18(3):339–353, 1971.

297. R. M. Smullyan. *Diagonalization and Self-Reference*. Number 27 in Oxford Logic Guides. Oxford Science Publications, 1994.

298. B. Souček and IRIS Group. *Dynamic, Genetic and Chaotic Programming*. John Wiley & Sons, 1992.

299. B. Steffen, C. Barry Jay, and M. Mendler. Compositional characterization of observable program properties. *Informatique Théorique et Applications*, 26(5):403–424, 1992.

300. W. J. Stewart, K. Atif, and B. Plateau. The numerical solution of stochastic automata network. *European Journal of Operation Research*, 86(3), 1995.

301. K. Sutner. Classifying circular cellular automata. *Physica D*, 45:386–395, 1990.

302. K. Svozil. Constructive chaos by cellular automata and possible sources of an arrow of time. *Physica D*, 45:420–427, 1990.

303. K. Svozil. *Randomness and Undecidability in Physics*. World Scientific, 1993.

304. K. Svozil. Halting probability amplitude of quantum computers. *Jnl. Universal Computer Science*, 1(3):201–204, 1995.

305. H. Takahashi. The maximum invariant set of an automaton system. *Information and Control*, 32:307–354, 1976.

306. A. Tarski. A lattice-theoretical fixpoint theorem and its applications. *Pacific Journal of Mathematics*, 5:285–309, 1955.

307. G. Troll. Formal language characterization of transitions to chaos of truncated horseshoes. Technical Report 13, SFB 288, T.U. Berlin, 1992.

308. G. Troll. Formal languages in dynamical systems. *Acta Univ. Carolinae, Math. et Phys.*, 34(2):117–134, 1993.

309. A. van de Mortel-Fronczak. *Models of Trace Theory Systems*. PhD thesis, T.U. Eindhoven, 1993.

310. M. Vellekoop and R. Berglund. On intervals, transitivity = chaos. *American Mathematical Monthly*, pages 353–355, April 1994.

311. P.-F. Verhulst. Recherches mathématiques sur la loi d'accroissement de la population. *Nouv. Mémoires de l'Académie Royale des Sciences et Belles Lettres de Bruxelles*, XVIII(8):1–38, 1845.

312. G. Y. Vichniac. Boolean derivatives on cellular automata. *Physica D*, 45:63–74, 1990.

313. B. von Karger and C. A. R. Hoare. Sequential calculus. *Information Processing Letters*, 53:123–130, 1995.

314. J. Von Neumann. *Theory of Self-Reproducing Automata*. University of Illinois Press, Urbana, 1966.

315. B. Voorhees. Division algorithm for cellular automata rules. *Complex Systems*, 4:587–597, 1990.

316. B. Voorhees. Determination of fixed points and shift cycles for nearest neighbor cellular automata. *Jnl. Stat. Phys.*, 66(5/6):1397–1414, 1992.

317. B. Voorhees. *Computational Analysis of One-Dimensional Cellular Automata*. World Scientific, 1996.

318. X. Wang. Period-doublings to chaos in a simple neural network: An analytical proof. *Complex Systems*, 5:425–441, 1991.

319. X. Wang. Discrete-time dynamics of coupled quasi-periodic and chaotic neural network oscillators. In *Proc. of IJCNN 92, Baltimore*, 1992.

320. X. Wang and E. K. Blum. Discrete-time versus continuous-time models of neural networks. *Journal of Computer and System Sciences*, 45(1):1–19, 1992.

321. X. Wang, E. K. Blum, and P. K. Leung. Analysis and simulation of synchronization in oscillatory neural networks. In *Proc. of CNS 92, San Francisco*, 1992.

322. Y. Wang and H. Xie. Grammatical complexity of unimodal maps with eventually periodic kneading sequences. *Nonlinearity*, 7:1419–1436, 1994.

323. K. Weihrauch. *Computability*, volume 9 of *EATCS Monographs on Theoretical Computer Science*. Springer-Verlag, 1987.

324. K. Weihrauch and U. Schreiber. Embedding metric spaces into CPO's. *Theoretical Computer Science*, 16:5–24, 1981.

325. K. R. Wicks. *Fractals and Hyperspaces*, volume 1492 of *Lecture Notes in Mathematics*. Springer-Verlag, 1991.

326. S. Wiggins. *Introduction to Applied Nonlinear Dynamical Systems and Chaos*. Springer-Verlag, 1990.

327. J. C. Willems. Paradigms and puzzles in the theory of dynamical systems. *IEEE Trans. Autom. Control*, 36(3):259–294, 1991.

328. R. F. Williams. Composition of contractions. *Bol. da Soc. Brasil. de Mat.*, 2(2):55–59, 1971.

329. G. Winskel. A compositional proof system on a category of labelled transition systems. *Information and Computation*, 87:2–57, 1990.

330. S. Wolfram. *Theory and Applications of Cellular Automata*. World Scientific, 1986.

331. S. Wolfram. *Cellular Automata and Complexity*. Addison-Wesley, 1994.

332. W. K. Wootters and C. G. Langton. Is there a sharp transition for deterministic cellular automata? *Physica D*, 45:95–104, 1990.

333. H. Xie. On formal languages in one-dimensional dynamical systems. *Nonlinearity*, 6:997–1007, 1993.
334. H. Xie. *Grammatical Complexity of One-Dimensional Dynamical Systems*. World Scientific, 1997.

Glossary of Symbols

(X, f) ... dynamical system, relation f defined on X

$\heartsuit$ .... undefined state
$\boxtimes$ .... end of example
$\square$ .... end of proof
$\because$ .... because (in proofs)
$\square$ .... always (temporal logic)
$\diamond$ .... eventually (idem)
$x \xrightarrow{f} y$ state transition of f, from x to y
$\rightsquigarrow\!\!\rightarrow$ .... attraction
$\rightsquigarrow\!\!\rightarrow_\omega$ .... infinite attraction
$\rightsquigarrow\!\!\rightarrow_\eta$ .... transfinite attraction
$\circ$ composition of composition operators
$\div$ .... product covering
$|$ .... such that
$\#$ .... cardinality
$!$ ..... Hilbert's nondeterministic choice
$\equiv_\rho$ .... ρ-equivalence
$=_\Pi$ distributivity of Cartesian products

$\leq_c$ .... computational power order
$<_c$ .... strict relation
$\equiv_c$ .... equivalence relation

$\emptyset$ .... empty set
$\subseteq$ .... set inclusion, inclusion-order
$\sqsubseteq$ .... definition-order
$\approx$ .... dense inclusion

$\Subset$ .... abstraction homomorphism

X^n .... sequences of length n
$X^{\leq n}$ .... sequences of length at most n
X^* .... finite sequences
$X^\omega, X^{\mathbb{N}}$ .. infinite sequences (length ω)
$X^{\mathbb{Z}}$ .... bi-infinite sequences
X^∞ .... finite or infinite sequences
$X^{n>\omega}$ .... strictly transfinite sequences
$X^{\mathbb{O}}$ .... transfinite sequences

$\uparrow$ .... increasing sequence
$\downarrow$ .... decreasing sequence

$\preceq$ .... partial order on words
$\prec$ .... strict order on words

s^+ .... positive sequence $s_0 s_1 s_2 \cdots$
s^- ..... negative sequence $s_0 s_{-1} s_{-2} \cdots$
$,\sigma$ .... future trace $\sigma_0 \sigma_1 \sigma_2 \cdots$
$\sigma,$ .... past trace $\cdots \sigma_2 \sigma_1 \sigma_0$

$\leq$ .... lattice order
$\sqcap$ .... greatest lower bound
$\sqcup$ .... least upper bound
$\bot$ .... bottom element
$\top$ .... top element

$\dot{f}$ .... derivative or gradient of f
$\frac{\partial f}{\partial x}$ .... partial derivative
$\tilde{x}$ .... U–u, D–d, V–Λ, 0–1
$\tilde{A}$ .... approximation of set A
$\overline{A}$ .... closure of A
$\underline{s}$ .... mirror image of sequence s
$\lceil x \rceil$ .. integer upper approximation of x
$[a, b]$ .... closed interval
(a, b) .... open interval
$|w|$ .... word size
$w|n$ .... length n prefix of a word

f^{-1} .... inversion
$(A \rightarrow f)$ .... domain restriction
$(f \leftarrow B)$ .... range restriction
$(f \rightarrow A \leftarrow B)$ ...... domain and range restriction
$\sim f$ .... negation
$\neg f$ .... external negation
$f \backslash g$ .... difference
$f; g$ .... sequential composition
$f \cap g$ .... intersection
$f \cup g$ .... union

$f \cup_N g$ .... nondeterministic union
$f \cup_F g$ ..... fair nondeterministic union
$f \cup_{P:p} g$ .... probabilistic union
$f \times g$ .... free product
$f \otimes_R g$ .... connected product
$\star$ .... generic composition operator
$\diamond$ .... generic property composition

d reverse down folding, generic distance
d_e .... Euclidean distance
d_a .... astronomer's metric
d_s .... Boolean distance
e .... label
f .... generic dynamical system
$\mathbf{f}$ .... Boolean "false"
g .... generic dynamical system
h .... Hausdorff metric
s .... generic trajectory
t .... idem
$\mathbf{t}$ .... Boolean "true"
u .... reverse up folding

A_i .... covering elements
D .... down folding action
E ... configuration space of the general computational model
G .... generic global property
H .... Hamming distance
H^p .... shifted Hamming distance
I .... generic individual property
J .... invariant
J_-^f .... greatest potential backward invariant
J_+^f greatest potential forward invariant
J^f .. greatest potential global invariant
J_f^- . least necessary backward invariant
J_f^+ ... least necessary forward invariant
J_f ..... least necessary global invariant
K .... Lipschitz coefficient
N_x .... neighborhood of x
P .... attracted set
Q .... attracting set
R ... neighborhood relation (connected product)
T .... topology
U .... up folding action
V .... valley profile
W .... word
X .... state space
Y .... state space

$\mathcal{A}$ .... complex CA
$\mathcal{B}$ .... bottom relation

$\mathcal{C}$ .... CA configuration space
$\mathcal{C}/_{\equiv_\rho}$ .... CA shift-quotient space
$\mathcal{D}$ .... cantor set
$\mathcal{E}$ .... empty relation
$\mathcal{F}$ .... fixpoint CA
$\mathcal{H}$ ..... energy-like decreasing function
$\mathcal{I}$ .... identity relation
$\mathcal{J}$ ...... set of folding instructions
$\mathcal{L}$ .... set of landscapes
$\mathcal{N}$ .... null CA
$\mathcal{O}$ .... algorithmic time complexity
$\mathcal{P}$ .... periodic CA
$\mathcal{R}$ .... set of relations
$\mathcal{S}$ .... shifting CA
$\mathcal{U}$ .... universal relation
$\mathcal{V}$ .... concrete observation function
$\mathcal{W}$ .... abstract observation function
$\mathcal{X}$ .... concrete label state
$\mathcal{Y}$ .... abstract label state
$\mathcal{Z}$ .... abstraction function

$\mathbb{A}$ .... invariance property
$\mathbb{B}$ .... set of Boolean values
$\mathbb{C}$ ... computable infinite configurations
$\mathbb{E}$ .... reachability property
$\mathbb{F}$ .... evolution operator
$\mathbb{G}f$ .... greatest fixpoint of f
$\mathbb{I}$ .... effect domain
$\mathbb{J}$ .... strongest invariant
$\mathbb{K}(X)$ nonempty compact subsets of X
$\mathbb{L}$ .... lattice
$\mathbb{N}$ .... natural numbers
$\mathbb{O}$ .... ordinal numbers
$\mathbb{O}_s$ .... successor ordinal numbers
$\mathbb{O}_l$ .... limit ordinal numbers
$\mathbb{P}(X)$ .... power set of X
$\mathbb{R}$ .... real numbers
$\mathbb{S}f$ .... least fixpoint of f
$\mathbb{U}$ uncomputable infinite configurations
$\mathbb{Z}$ .... integers

α .... covering
γ .... contractivity factor
δ .... infinitesimal value
ε ...... empty word, infinitesimal value
ζ .... TM transition function
$\eta_\mathbb{L}$ .... cardinality of lattice $\mathbb{L}$
θ .... point-level forward dynamics
κ .... kind
λ ...... parameter (logistic map, Smale horseshoe, Langton)
μ .... parameter (Smale horseshoe)
ξ .... set-level forward dynamics

π .... trace-state function
ρ .... shift
σ .... trace
τ .... trace
υ .... threshold function of H^ρ
ϕ .... encoding relation (simulation)
χ .... instruction-landscape function
ψ .... decoding relation (simulation)
ω .... first transfinite ordinal

Γ .... functional
Δ .... action domain
Θ .... point-level dynamics
Λ .... ridge profile
Ξ .... set-level dynamics
Π .... projection
Σ .... alphabet
Υ .... dynamical complexity
Φ ... set of approximating sets (fullness and atomicity)
Ψ .... (idem)

Index